Sustainability and Environmental Economics
An Alternative Text

Sustainability and Environmental Economics:
An Alternative Text

John Bowers

 Longman

Addison Wesley Longman

Addison Wesley Longman Limited,
Edingburgh Gate, Harlow,
Essex CM20 2JE, England
and associated companies throughout the world

First published 1997

British Library Cataloguing in Publication Data
A catalogue entry for this title is available from the British Library

ISBN 0-582-27656-X

Library of Congress Cataloging-in-Publication Data
A catalog entry for this title is available from the Library of Congress

Set by 30
Produced by Longman Singapore Publishers (Ptc) Ltd
Printed in Singapore

Contents

Preface

My work as an environmental economist has long had two aspects: the purely academic world of seminars and lectures, examiners' meetings and administrative rows; and the practical world of environmental conservation, composed of public inquiries, responses to discussion documents, evidence to parliamentary committees and crisis meetings on environmental threats. Traditionally the academic teaches during terms and researches and consults during the vacations. Increasingly, as terms grow into semesters and teaching duties grow exponentially, research and consultation occurs in whatever spare moments can be found. It is still possible to occupy both worlds, but only just.

The ideas presented in this book have grown out of the interplay between these two worlds and I have benefited from meeting and working with people in both.

Among academic economists my debt to Claude Henry and Michael Young is fully and gratefully acknowledged. I owe a debt also to Chris Nash and Quentin Outram who listened to my ideas and gave helpful comments. Academics, but not economists, who have made a contribution, although they may be surprised to hear it, are Graham Cox, Phillip Lowe and Sarah Whatmore. Finally, among academics, I owe a special debt to Peter Hopkinson who shared in some of the research and encouraged the writing of this book.

I have enjoyed working with a large number of conservationists and many ideas have developed out of the work we did. I owe particular debts to Simon Bilsborough, Rick Minter and Gwyn Williams. I am especially grateful to Stephen Warburton in whose company I have fought many skirmishes and a few battles.

But there are not two worlds but three. My family have supported, encouraged, tolerated (just), cajoled and criticized me before, during and after the book's conception and production. It is dedicated to Anne, Allison and Simon.

Part 1

Introduction

Chapter 1

Environmental problems and environmental economics

Economic activity affects the environment in many diverse ways. In producing and consuming goods and services societies modify the chemical compositions of the atmosphere, soils, fresh waters and oceans; alter the vegetation cover of the land and the diversity of wildlife inhabiting both land and water. Some of these environmental modifications are intentional, such as those achieved through processes of agriculture and forestry, from urbanization and the construction of social infrastructure, e.g. roads, factories and power stations. Other environmental impacts are incidental and often unintentional by-products of economic activity. These include discharge of wastes from industry and domestic living, impacts of tourism and spillover effects of urbanization.

Current concerns with the environment, what has been called the environmental crisis, rest on the proposition that environmental modification has gone too far; to the point where the welfare of current and future populations is adversely affected; and that therefore both intentional and incidental impacts of economic activity on the environment have to be reduced and ameliorated. This view underlies the notion of sustainable development to which, following the Earth Summit in Rio, almost all the nations of the world are committed.

In discussions of sustainable development the famous definition of the Brundtland Commission of 'development that meets the needs of the present without compromising the ability of future generations to meet their own needs' (World Commission on Economic Development, 1987) is much quoted. This definition has found ready acceptance because of its generality. To make it operational requires interpretation and specification, and in that process much complexity and confusion has arisen. Some of the fundamental questions that arise are as follows:

1. Is sustainable development compatible with continuing economic growth?
2. Is it compatible with continuing population growth at the rates currently being experienced?
3. Does it require sacrifices on the part of current generations, or simply modifications in the ways in which they conduct their affairs?
4. How are the costs of sustainable development to be distributed between nations and, within each nation, how are the costs to be distributed between different strata of society?

5. Does sustainable development imply that Third World countries will be denied the opportunities to achieve the levels of per capita income experienced by countries in the West?
6. What are the resources that future generations will need?
7. How will we decide whether current activities constitute (are compatible with) sustainable development?
8. What policy measures have to be adopted in order to achieve sustainable development?

Economics cannot on its own provide a full answer to any of these questions, each of which involves broad social (and moral) issues that can only be resolved, if at all, by extended and wide-ranging debate. However, economics has a contribution to make to each of them. Thus on question 1 economists have identified possible definitions of sustainable development and shown formally how sustainable development using these various definitions differs from the objective of maximizing the rate of economic growth. Question 2 involves the moral issue of population policy on which the world is deeply divided; economics can illustrate the resource implications of different population growth paths. Population growth and its relationship with exhaustible resources is also at the root of question 5. On question 3, there has been extensive work on the implications for national incomes of meeting specific sustainability objectives, notably controlling global climate change. On question 4 the principal role of economics is in revealing the implications for income distribution of alternative policy measures to achieve sustainability objectives. The issues underlying question 6 also go beyond economics. Economists have attempted to provide complete answers only to the last two questions in this list and the thinking on them is discussed later in this book.

In summary, economics has three important roles in debates on sustainable development:

- To examine the costs and benefits of achieving particular sustainability objectives, such as limiting the emissions of greenhouse gases or conserving biodiversity.
- To assess the effectiveness of alternative policy instruments for meeting those objectives. An understanding of economics is required in order to predict how people will behave when faced with different types of taxes and subsidies or legally enforced environmental standards.
- To assess the costs and benefits of alternative policy instruments.

Environmental economists aim to put a monetary value on the environmental effects of economic decisions and to provide a framework for comparing the environmental losses with economic gains. In doing this they draw on the body of theory and doctrine of environmental economics. The object of this book is to explain what that theory is and to evaluate the role of economics in resolving the environmental crisis. It presents the tools of analysis and the current doctrines of economics on sustainable development and environmental problems in general.

Environmental economics applies economic analysis to problems of the environment. Environmental economists seek to understand the economic causes of environmental problems and to use that understanding to specify policies to treat these problems. The conventional economic analysis used is known as neoclassical microeconomics. This is essentially the theory of the operation of markets. It explains how markets distribute resources and states the conditions under which freely functioning markets will maximize society's welfare. In the neoclassical view environmental problems arise from failures of markets to properly allocate resources and the role of the economist is to define measures to correct those failures and allow markets to generate some best or optimum state of the environment. Thus, if economic activity is not sustainable, then this is because markets (economists often speak loosely of *the* market, meaning the market economy) are failing to make adequate provision for the future. Equally if a river is being polluted by a factory, then this is because the market is failing to ensure that the factory owner pays for the damage that is being caused. In each case diagnosis of a failure of the market leads to proposals for remedying that failure, i.e. making markets take account of the needs of future generations, or making polluters pay for their pollution. In extreme cases the market failure may be impossible to cure and the market must be replaced by some other means of social control, such as the criminal law. But neoclassical economics teaches that markets are a uniquely powerful and efficient device for allocating resources and neoclassical economists are reluctant to consider non-market solutions.

Concern with environmental problems has led to the development of alternative schools of economic thought. One such school is ecological economics, which attempts to integrate ecology with economics. Others have considered the implications of the laws of physics, particularly entropy, for the limits to economic activity. There is also the so-called New Economics, which advocates self-sufficiency, small-scale enterprise and the severe modification if not abandonment of industrial societies. But where these alternative schools of thought consider economic behaviour, the theory they use is still neoclassical and an understanding of the neoclassical theory of environmental economics is therefore necessary to a proper understanding of them, if only because neoclassical views are their point of departure.

One function of this book is to explain conventional neoclassical analysis of the environment. But it also contains a critique of it and argues that the conventional wisdom needs modifying in a number of important respects. What it argues can be briefly summarized under the following headings. I express the conventional view as a series of propositions and summarize my criticisms of them.

The choice of instruments for environmental policy _____

Proposition

Environmental pollution and, by analogy, other environmental problems, result from market failure, the failure of markets to properly allocate resources. The role of the economist is that of the market doctor; she diagnoses these failures and prescribes treatment, designing policy instruments to correct the failures and restore the environment to the state that it would be in were the market able to function properly.

Criticism

For the serious forms of pollution that provide the *raison d'être* for environmental economics, this vision does not fit. Environmental problems are discovered and become objects of social concern that need to be dealt with. They are more akin to new goods than to inefficiently produced old ones. The market has not failed but there are new problems that it does not cope with. It may be that the market can be modified to accommodate these new concerns but they may require solutions outside of the market with market forces constrained as a consequence.

Proposition

There is an optimum level of pollution or state of the environment which is what would occur if market failures were corrected. It depends on the value placed by the sufferers on the damage that they suffer and the benefits that polluters receive in the process of causing the pollution.

Criticism

Since the market has not failed there is not an optimum level of pollution which a successful market would generate. Faced with serious decisions about the state of the environment, conscious collective choices have to be made. These cannot be left to the dictates of the market, nor to witch doctors offering to read the market's entrails. In economic terms the environment is a collective good that requires collective democratic decisions. Democracy cannot be replaced by one of its artefacts, the market.

Proposition

Since polluters are causing damage to others it is appropriate that policies to correct the problem should make the polluter pay.

Criticism

Because many forms of pollution, and all the serious ones, do not result from market failure there is no presumption that the polluter should pay. If society discovers, say, global warming, and decides that in order to control it emissions of CO_2 should be reduced, it is faced with issues of efficiency, i.e. how to meet its objectives at least cost, and with equity, i.e. how to distribute the resulting burden. Efficiency does not dictate that the polluter should pay and it is not obviously equitable that she should do so. In the presence of irreversibility the polluter-pays principle is not compatible with efficiency.

Proposition

Economic or market instruments are more efficient than administrative and legal controls in dealing with environmental problems and particularly with pollution. They achieve the objective at lowest resource cost and additionally stimulate innovation in techniques for reducing pollution.

Criticism

Since pollution does not result from market failure there is no presumption that market instruments should be used to reduce it. Market instruments are allocatively efficient but that is only one consideration in the choice of instrument and it can be outweighed by other factors such as the cost of monitoring and enforcement. For several reasons control costs are often lower with administrative controls. While there are well-discussed cases where the efficiency gains of marketable permits outweigh all other considerations, generalization from these cases would be dangerous. The view that market instruments are unique in stimulating innovation in pollution-control techniques is false.

Cost–benefit analysis

Proposition

The environment is damaged in public-investment decisions because environmental effects are unpriced. The optimum balance between environmental conservation and public investment would be achieved if all environmental effects were given monetary valuation and incorporated into cost–benefit analysis. Techniques now exist to do this.

Criticisms

While survey techniques permit monetary values to be given to environmental effects of investment programmes, these values have no clear meaning and do not approximate to some 'true' valuation. Indeed where the environmental assets are not traded in markets and therefore do not have current uses, they have no true market value at all. The problem is not that of market failure which may be rectified through the construction of surrogate markets. Markets do not and cannot exist for these things. To suppose otherwise is to misunderstand the nature of markets.

Even if it were possible to place meaningful monetary values on all environmental effects of projects, this would not ensure that the environment was safeguarded. This is because the textbook notion that cost–benefit analysis informs the independent decision-maker of where the social interest lies has no counterpart in reality. Cost–benefit analysis is used strategically by public-sector agencies to achieve their objectives. These objectives exist independently of the balance between costs and benefits. Their response to a requirement to value environmental effects could be expected to be a strategic one, which would neutralize the effects of monetary valuation.

The effective way to protect the environment in public-sector decision-making is through the use of environmental standards as constraints on the decision processes. The standards imposed reflect society's desires about the protection of the environment. They are in no sense approximations to what some hypothetical set of well-functioning markets would produce.

Sustainable development

Proposition

There are two concepts of sustainability: weak sustainability, which is satisfied if losses of natural capital are compensated for by increases in man-made capital of equal value, and strong sustainability, which requires that aggregate natural capital does not decrease.

Criticisms

Both concepts depend on the notion of aggregate natural capital. This notion requires that disparate entities (e.g. the capacity of the atmosphere to sustain life and the quantity of biodiversity) can be measured in comparable units aggregated so that judgements can be made as to whether aggregate capital has risen or fallen between different situations and, if it has fallen, whether changes in man-made capital compensate for the losses. The comparable units are, of course, money values and the neoclassical approach to sustainable development rests on the capacity to place money values on the environment. The rules for sustainable development, then, are modifications to the rules of cost–benefit analysis. This approach is rejected for reasons given in the previous section.

Proposition

Sustainable development is the maximum rate of growth or development that meets the above sustainability constraints.

Criticisms

Sustainable development is growth or development subject to a set of specific environmental standards imposed on decision-making. These standards should be constructed to meet perceived requirements for sustainability: safeguarding the atmosphere, protecting biodiversity and so on. They cannot be traded off against each other and cannot be expressed in terms of money valuations or concepts such as natural capital.

Proposition

A commitment to sustainable development does not alter conclusions about the superiority of economic instruments over administrative and legal controls; indeed it gives added emphasis to their importance. However the risks of irreversible extinctions of plants and animals and of irreversible changes in the earth's atmosphere may mean that administrative controls and legal sanctions are needed as a safety net to back up economic instruments.

Criticisms

Sustainable development introduces new considerations into instrument choice. It introduces an international dimension, which will appear as agreed targets for reduction in global pollution, etc. and these should be embodied in environmental standards. Since decisions can be irreversible, avoiding failure is an important consideration and this changes the traditional debate. Both economic and administrative and legal means of achieving objectives are prone to fail. Legal sanctions do not provide a usable back-up and safety net against irreversibility. Sustainable development therefore requires a new approach to achieving environmental objectives. Individuals have to accept the objectives and see sustainability as being to their advantage. Economic activity that is compatible with sustainability and which gives individuals property rights in the environment is what is required.

Structure of the book

To understand the issues in environmental economics requires some basic knowledge of neoclassical theory of markets and market behaviour. This is the subject of Chapter 2 and readers who have never studied economics should read it. Those who have previously studied economics can bypass it without loss. Chapter 2 presents the argument by use of graphical techniques, which are used in subsequent chapters in preference to algebra. Students new to economics tend to find graphics difficult but the effort of understanding is worthwhile. Economic arguments are of themselves difficult, involving several variables, and graphics are a way of presenting them simply and clearly.

Chapter 3 is devoted to the notion of market failure, which is central to environmental economics and the concepts are used extensively in succeeding chapters. Again, readers with previous knowledge of economics may skip it and in any case can refer back to it as necessary.

The body of environmental economics is grouped under three broad sections: the choice of instruments for environmental policy, cost–benefit analysis and sustainable development.

The choice of instruments for environmental policy covers the analysis of environmental problems as externalities and the design of policy instruments to deal with them. These are central topics in environmental economics, which is concerned with how environmental problems arise and policies for dealing with them. The concentration is on pollution control. Pollution is a central issue in environmental protection and many problems not strictly concerned with pollution are treated by environmental economists as analogous to pollution. The issues covered here are central to an understanding of the economist's views of sustainability. They include the idea of an optimum state of the environment and the case for the use of economic methods of control as opposed to administrative and legal ones. The simple theory relates to situations where the number and location of polluters is known. Real-life problems are often more complex and the section contains case studies of three forms of diffuse pollution: nitrates in water, domestic waste and road transport emissions.

Cost–benefit analysis is a major tool of environmental economics and central to discussion of sustainable development policies. This section covers cost–benefit analysis in theory and practice with case studies of its use in the roads programme and land drainage schemes, two areas of environmental controversy. Placing money values on the environmental effects of economic development is a central activity of environmental economics. Neoclassical views of sustainable development require monetary valuation in cost–benefit analysis. This important matter is given a separate chapter.

All three sections bear on the issue of sustainable development. The third section examines in greater detail the meaning of sustainable development and the notions of strong and weak sustainability. It also considers in detail the economics of international cooperation in dealing with global, 'transboundary' pollution, policies for the conservation of biodiversity and problems of exhaustible and renewable resources.

Chapter 2

The economist's view of the world

This chapter covers the basic economic theory used in the succeeding chapters of the book. It is intended for the reader who has not studied economics and knows nothing of its doctrines. The reader who is familiar with the material may skip it without detriment to understanding the rest of the text.

Economics is conventionally divided into microeconomics and macroeconomics. Macroeconomics is the study of whole economies and concerns broad economy-wide economic phenomena such as the rate of inflation and the level of unemployment. Macroeconomic theories incorporate within them assumptions about the economic behaviour of individual economic units, individuals, households and business firms. The study of the economic behaviour of these individual economic units is the task of microeconomics. More formally, microeconomics can be defined as the study of resource allocation decisions under scarcity. It defines efficiency in resource allocation and investigates the conditions to achieve it. Neoclassical economics seeks to show that efficiency in resource allocation is achieved through competition in market economies. This chapter explains what that definition means. By the end of it the reader should know what is meant by resources and by resource allocation decisions. He or she should also understand the concept of scarcity and its importance, know what is meant by efficiency in resource allocation and understand why competitive markets achieve it. For subsequent reference these and other technical terms are defined in a glossary at the end of the chapter.

Some essential preliminaries

The essential basic concepts are best explained through a simple model of an isolated individual social unit, an individual family unit, which we will call a household, engaged in subsistence activity. Complications arising from the social structure will be dealt with subsequently.

The household has a number of needs: food, clothing, fuel, etc. In satisfying these needs it is engaging in economic activity. The object of its economic activity, meeting its needs, is defined as *consumption*. This term is used to encompass much more than the ingestion of food to include all the *goods* and *services* that individuals and households acquire and use (services are things like haircuts as opposed to tangible objects of consumption like cheese, shirts and stuffed redshanks; we will speak here as though households only consume goods). In this simple model the household is assumed to *produce* all these goods (and services) using the *resources* at its disposal. These resources are of course many and varied but are placed by economists into three broad classes termed the factors of production: land, labour and capital.

Land is defined as the physical space in which activities take place but may also be interpreted to encompass the raw materials or the 'products of nature'. For present purposes we may ignore it and concentrate on the other two.

Labour is the time available for the production of goods by the members of the household. That time is of course limited by basic biological needs of sleep, plumage maintenance, eating, etc., i.e. consumption requires time that is thus not available for labour.

Capital comprises the tools and machines at the household's disposal. This capital will be composed of a number of diverse items. An individual item is called a *capital good*. The essential feature of capital is that it has to be produced using these factors of production, but it has been produced in the past, i.e. capital is the result of past production utilizing labour, land and capital. Thus capital requires tool making as well as tool using. There are a number of animals that use tools, for instance Egyptian vultures use stones to break the eggs of ostriches. But the vulture's stone is not capital to the economist. Humans are postulated to be tool makers and the essence of tool making is that it is carried out before the specific act of use of that tool is known. If the vulture were to spend time finding stones that it carried with it on the chance that it would find an egg, then those stones would be capital in the economist's sense, since they would be the result of a prior commitment of labour.

At any time, given the resources available to it, the household is faced with a choice between devoting those resources to the production of goods for consumption or to the production of capital goods. The resources devoted to production of capital is termed the households *savings*. Savings therefore increases the quantity of capital available in the future.

At any time the household's consumption is constrained by the resources available to it. If it had more resources it would engage in more consumption and more savings. This is the *principle of scarcity*. Resources are scarce relative to the uses to which they can be put.

It follows from this that any commitment of resources to the production of any particular consumption or capital good has an *opportunity cost* in terms of the opportunity forgone by that decision. Given the constraint implied by the principle of scarcity any decision to commit resources to one line of production reduces the resources available for other purposes. Hence the opportunity cost of that decision is the production it precludes.

The notion of opportunity cost is equally applicable to animals other than humans provided that the animal faces a time constraint on satisfying its needs. Thus in studies of foraging strategies of redshanks (a species of wading bird) on tidal feeding grounds, ecologists have discovered that the distribution of available food items across the feeding area varies with the state of the tide. At a given state of the tide more food can be found for a given amount of foraging in some parts than in others. Birds are faced with a range of possible feeding strategies involving moving around the feeding area and foraging in different parts of it over the tidal cycle. The opportunity cost of a chosen strategy is the additional food that would have been yielded by an alternative strategy. From this one has the notion of an optimum strategy. This is the strategy that yields the maximum return from foraging activity. If food is scarce and is the factor determining the survival of the animal, the optimum strategy maximizes the probability that the bird does not starve. This is often the crucial factor in the winter months. Alternatively if food is abundant and breeding activity is making demands on the bird's time, the optimum strategy minimizes the time devoted to feeding and maximizes the time available for other activities. In this case the opportunity cost of a suboptimal strategy is forgone time for breeding activities.

The redshank, at least as I have described it, is a fairly simple animal. It is a hunter gatherer. Unlike humans it is not a tool user nor is it a social animal. Time is its only factor of production and there are relatively few uses to which it puts its time. Its resource allocation problem is a relatively simple one. Time, equally, is the root source of opportunity cost for humans but even with an isolated subsistence household the problem is very much more complex. It is perhaps worth spelling out the reasons why this is so:

- The range of uses of time is much greater because there are many more forms of consumption.
- Just as the redshank has a number of alternative strategies for acquiring its food, the household will have a number of alternative ways of producing its goods.
- Production (and consumption) are social activities entailing cooperation between individuals.
- Because of tool using there are other resources (factors of production) to be considered.
- Decisions have to be taken between the allocation of time for consumption activities and for savings, i.e. increasing the quantities of artefacts (capital) for use in future periods.

The concern of neoclassical microeconomics is to define efficiency in resource allocation, the counterpart of the optimum foraging strategy in the model of the behaviour of the redshank.

Techniques of production

Assume, in this simple model of the subsistence household, that the production of any good, whether for consumption or a capital good, requires the use of quantities of all factors of production, i.e. production involves both labour time from the household and capital goods. Obviously different forms of production will involve different capital goods and may involve different amounts of time of individual household members but if the factors are defined broadly then the assumption is reasonable.

A particular mixture or combination of factors for the production of a good is defined as a *technique of production*. There will typically be more than one technique for the production of each good. It is normally assumed also that the available techniques tend to increase over time. This phenomenon is known as technical progress. In the broad sweep of history technical regress is also possible since technology is culturally determined. Thus the technology of the Roman empire, which permitted the existence of large cities, was lost during the Middle Ages. However modern economics, which has been developed since the Industrial Revolution, admits only of technical progress.

Techniques differ in the quantities of factors of production that they require. Some techniques will be labour intensive, involving much labour and little capital; others will involve more capital but little labour. It is useful to assume that both labour-intensive and capital-intensive techniques are available for the production of each good. Thus the land may be dug by hand or ploughed with a horse. Hand digging is very time consuming but uses only a simple spade, which takes little time to make or to replace when broken. Horse ploughing is much quicker but a plough is a much more elaborate piece of equipment that takes longer to make and, in addition, oats have to be grown to feed the horse, which is time consuming.

A technique that uses more of all factors of production will never be used because, from the principle of scarcity, the household, by using an alternative technique, could consume that good and have available more scarce resources for the production of other goods. Techniques requiring more of all factors of production are termed *inefficient techniques*. Only *efficient techniques* are relevant to the household's decision problem. Consider a third means of tilling the land, by use of an ox plough and two oxen. The ox plough is heavier and takes longer to produce than the horse plough and two oxen eat more oats than a horse. If the oxen walk more slowly than the horse so that it takes longer to till the land with them than with the horse or it takes the same time but two people are needed, one to lead them and one to work the plough, while only one is needed with the horse, then ox ploughing is an inefficient technique and the household would not use it. If it used the horse then it would release labour time for other activities and would have to devote less of its savings time also to growing fodder and making ploughs. On the other hand, digging is not an inefficient technique, since although it involves more labour time in tilling less time is required to produce the tools and land is not needed for fodder production. If the household was short of land or lacked the resources to make a plough it would dig the land by hand. There are, however, no circumstances under which it would choose ox ploughing.

A technique that is efficient at one time may subsequently become inefficient because of technical progress. Ox ploughing will not be inefficient if the household does not have access to a horse.

The opportunity cost of shifting from an inefficient technique to an efficient one will be negative: the household can produce the same quantity of goods utilizing less of its available resources thereby leaving resources for other forms of production.

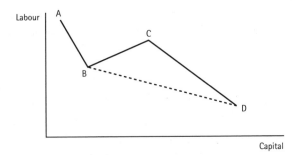

Fig. 2.1 Efficient and inefficient techniques.

This is shown in Fig. 2.1 where four alternative techniques for producing a given quantity of a good are plotted. Techniques A, B and D are efficient in that there is no technique that uses more of both factors of production. Technique C, however, is equivalent to the example of ox ploughing; it is inefficient and would never be used, since production with some combination of techniques B and D would use less of both factors of production (shown by the dotted line). Where the lines joining the techniques are positively sloped the technique is inefficient. Technique A is labour intensive, using relatively large amounts of labour time and relatively little capital. Technique D on the other hand is a capital-intensive technique. The household's choice of technique (between A, B and D) depends on the quantity of each resource it possesses. The more capital it possesses the lower is its opportunity cost relative to labour.

The set of all efficient techniques available at any time for the production of a given good, the recipe book if you like, is called the production function. In taking its resource allocation decisions the household will choose from its production functions for the various goods that it wishes to produce. Which technique it chooses will depend on the opportunity cost of the factors used. Where capital is scarce and labour abundant, labour-intensive techniques will be chosen.

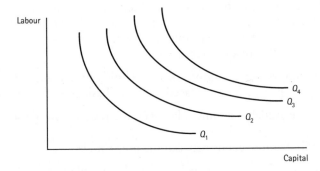

Fig. 2.2 A neoclassical production function.

Figure 2.2 shows a production function in what is known as the neoclassical form, where the number of available techniques is unspecified with continuous substitution permitted between factors of production. This makes no sense for the isolated subsistence household but may be treated as a convenient simplifying assumption for a whole complex society. Each curve corresponds to techniques available for a given level of output of the good in question. Each one is equivalent to what is shown in Fig. 2.1 and is generally called an *isoquant*. Isoquants comprise the set of efficient techniques for producing a given output of the good. Isoquants further from the origin correspond to higher levels of output of the good. Isoquants cannot intersect since they only relate to efficient techniques.

Resource use in society

If we abandon the assumption of the individual isolated household and think of societies composed of numbers of interacting households then we can introduce some additional concepts.

In economic activities humans engage in specialization and exchange goods with others. The exchange of goods is known as *trade*. Trade is dependent on specialization or what is known as the *division of labour*. The division of labour permits a society to increase its total production of goods and services for two reasons:

1. An individual specializing in an activity becomes more skilled at it. Hence a division of labour increases productivity (output per unit of input) including per unit of time devoted to that activity. The productivity of the foraging redshank is naturally measured as biomass of food items collected and consumed per unit of foraging time. For a tool user and especially for a tool maker, the notion of productivity becomes more complex because of the variations in the quantity of capital that is used.
2. The individual saves time in moving between tasks. The cost of division of labour is that the household no longer produces all of its needs. Hence the division of labour necessitates trade or exchange of goods and services. The institution of exchange is defined as the *market*. The division of labour is limited by the opportunities for exchange. This is usually expressed by the famous dictum of the eighteenth-century economist Adam Smith, that the division of labour is limited by the extent of the market. The market allows for the production of a greater range and volume of goods than is obtainable in a non-market or subsistence economy.

An economy comprising a number of households practising a division of labour and engaging in exchange is faced with two basic sets of decisions:

- how to allocate available resources between the production of the available goods;
- how to distribute those goods between the households.

From the principle of scarcity it may be inferred that only resource allocations and consequent distributions that are efficient will be acceptable. We have already encountered the notion of efficiency in the selection of techniques for the production of an individual product. This notion must now be extended to the problems of allocation and distribution. In essence efficiency requires that *opportunity cost is positive for all choices.*

Specialization and production efficiency

The problem at its simplest can be presented if we assume an economy comprising a number of cooperating households that wishes to produce two goods, which we will call wheat and cloth. The allocation problem can be illustrated through a construction known as an *Edgeworth box*. If the dimensions of the box represent the available quantities of resources and the origin of the production functions for the two goods are fixed at opposite corners, then the box represents the set of feasible allocations of the resources between the production of the two goods. Figure 2.3 shows the Edgeworth box for our simple economy. The quantity of capital available to it is represented by the horizontal dimension of the box and the quantity of labour by the vertical dimension. The origin of the production function for wheat is the bottom left-hand corner and the origin of the cloth production function is the top right-hand corner. Take a point in the box such as *T*. The co ordinates of *T*, when measured from the bottom left-hand corner, give quantities of labour and capital allocated to the production of wheat. The remainder of the available factors would then be allocated to cloth and those amounts are the coordinates of *T* when measured from the top right-hand corner.

A subset of these allocations, where the contours of the two production functions are tangential, is the *contract curve. Allocations in this subset are efficient in the sense that, between them, opportunity cost is positive*; i.e. more wheat may only be produced by producing less cloth. Allocations off the contract curve will not be efficient. It will be possible, by moving to the contract curve, to produce more wheat without the sacrifice of the production of cloth. Thus opportunity cost will be negative for allocations off the contract curve. From the principle of scarcity the households therefore will desire to locate on the contract curve.

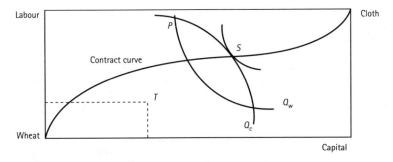

Fig. 2.3 The Edgeworth production box.

This can be illustrated by taking another point *P* in Fig. 2.3. *P*, like *T*, is a feasible allocation of factors of production between wheat and cloth. It is a point of intersection of isoquants of the two products (denoted as Q_w and Q_c) and as such it is an inefficient allocation, since any allocation within the area bounded by the two isoquants will allow the production of more of both goods. An efficient point is shown, *S*, where the two isoquants are tangential. It is then only possible to produce more wheat, by reallocating resources, at the expense of production of cloth. Thus if we start on the contract curve at *S* the opportunity cost of producing more wheat is a reduction in the production of cloth, but from *P* the opportunity cost in terms of forgone cloth from producing wheat is negative. By moving to the contract curve we can have more of both products.

The various combinations of goods that would be produced at points on the contract curve represent the outer boundary of production possibilities for the household. If the contract curve is portrayed in terms of the goods produced it is known as the *production possibility frontier* (PPF). This is shown in Fig. 2.4. On the PPF, opportunity costs are positive and are given by the slope of the frontier (the opportunity cost of producing an additional unit of wheat is a number of units of cloth). Allocations outside of the PPF are unobtainable. Allocations within the PPF have negative opportunity costs (it is possible to produce more of both products). From the principle of scarcity the economy will wish to allocate its resources so as to locate on its PPF.

Consumer choice theory

In our discussion of the time-allocation problem of the redshank, we found it necessary to have some understanding of its preferences in order to specify an efficient allocation. Our model was a very simple one: at critical times in the winter when food was scarce, finding enough to eat dominated its preferences; at other times it had a high preference for non-foraging time in order to improve its breeding success. With human societies our model is, of necessity, more complex. None the less it is still true that in order to answer the question

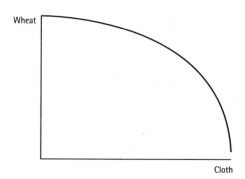

Fig. 2.4 The production possibility frontier.

concerning distribution of the goods produced we need to introduce the notion of household (consumer) preference.

Households will have a preference function for the goods they wish to consume. They will prefer combinations offering more of all available goods to combinations offering less but in general will be willing to trade off quantities of some goods for quantities of others. Between combinations offering more of some but less of others the choice is thus less clear. We may postulate combinations between which the household is indifferent and envisage the preference function as comprising contours of such combinations. These contours must have negative slopes and will in general be convex when viewed from the origin (like the contours of the production function), implying a declining willingness to sacrifice one good to acquire another as the quantity of the former good decreases. The scarcity postulate then leads us to define the household's objective as using its resources and its trading opportunities to reach the highest possible contour of its preference function. An illustration of a preference function for a household faced with just two goods (again for simplification and diagrammatic exposition) is shown in Fig. 2.5. The household will prefer to be on contour 6 to contour 4 since on 6 it can have more cloth and more wheat but will be indifferent to different positions on a contour. Along a contour it 'trades off' one good for the other.

With these assumptions we can apply the concept of efficiency to exchange or trade. Diagrammatic presentation is for two households, A and B and the two goods wheat and cloth. We again construct an Edgeworth box (Fig. 2.6), this time with dimensions representing the available goods for distribution and with the two consumers located at opposite corners. Each point in the box is a feasible allocation between the households and by assumption each allocation puts each household on some contour of its preference function. *P* is such a point and since it is off the contract curve the contours of the preference functions intersect. As before, the contract curve, comprising points of tangency between contours of the two preference functions, is the subset of efficient distributions. Since *P* is off the contract curve then both households can benefit from exchanging goods and, if initially allocated goods so as to place them at *P*, will be motivated to do so. At any point in the area between the two contours they will both be better off. It is said that at *P* there are *gains from trade*.

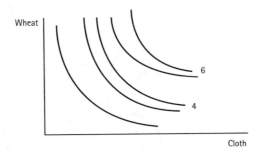

Fig. 2.5 A household's preference function.

But at what rates of exchange will trade take place? The slopes of the lines from *P* to *Q* and *R* represent exchange rates for goods. The trade of wheat for cloth, which will move the households from *P* to *Q*, is given by the tangent to angle *e*. Trade at a rate equal to the tangent to angle *f* will move them from *P* to *R*. These are the limits to acceptable trades since at *Q* household A is on the same contour of its preference function as at *P*, and at *R* household B is on the same contour as at *P*. For any exchange rate between these two extremes there are gains from trade and households will wish to trade. On the contract curve gains from trade are exhausted. One household can only be made better off at the expense of the other. At *P* both can be better off by exchanging goods.

Efficiency and welfare

We have now established two conditions that are necessary if an economy is to allocate its resources efficiently:

1. it must be on its PPF;
2. gains from trade must be exhausted.

These conditions are not unique of course: there are many allocations that are on the PPF and gains from trade are exhausted at any point on the contract curve of the Edgeworth trading box.

Where we reached on the contract curve depended on our initial arbitrary allocation of goods to point *P*. An alternative allocation, for instance to point *T* or point *U* in the trading box, would have opened up different segments of the contract curve to gains from trade. Clearly the individual households will not be indifferent to the starting point. Household A would prefer to start at *T* rather than *P*, since this would place it on a higher preference contour. For the same reason, household B would prefer *U* to *P*. Thus, in large part, the welfare of individual households depends on how goods are initially allocated. Given this initial allocation, trade can make them better off, but if household A receives the allocation of goods of *P*, it cannot, through trade, raise its welfare to the level that it would have achieved, without trade, had the initial

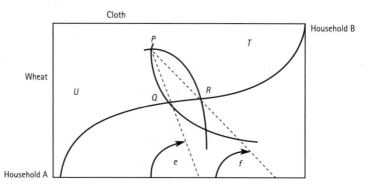

Fig. 2.6 The Edgeworth trading box.

allocation been at T. The process that determines the initial allocation is therefore important. In a capitalist economy this is determined by the household's possession of factors of production: its endowment of labour, but especially its ownership of capital. The greater the quantity of factors that the household brings to the cooperative production process, the greater its share of the fruits of that production.

The other factor that determines the welfare of individual households is the terms under which trade actually takes place. In Fig. 2.6 we saw that households allocated goods at P would be willing to exchange goods, to trade, at any exchange rates between the limits of tangents to the angles e and f. The former, moving the households to Q, would be preferred by B since that maximizes its gains from trade while leaving A neither better nor worse off than at P; i.e. at this exchange rate B captures all the gains from trade and A is indifferent. Conversely exchange at the rate tan f is preferred by A since it then captures all the gains from trade. The exchange rate determines how much of one of the goods the household has to give up to acquire a unit of the other good. In moving from P to Q or R, household A is giving up wheat to obtain more cloth. Conversely B is giving up cloth in order to obtain more wheat. Tan f means that more cloth is obtained per unit of wheat and is consequently preferred by A. Tan e is an exchange rate more favourable to household B, which is selling cloth and buying wheat.

Figure 2.6 was constructed for a specific combination of wheat and cloth produced by society and measured by the dimensions of the Edgeworth box. This could correspond to a single point on the PPF. We have seen that the condition for efficiency, that gains from trade are exhausted, can be satisfied by a number of points on the contract curve: segment QR if the initial allocation of goods starts us at P, and other segments if the allocation puts us at some point other than P. Any other point on the PPF will give us another Edgeworth box of different dimensions, i.e. more wheat and less cloth or more cloth and less wheat. Since the PPF is the locus of efficient production levels, alternative Edgeworth boxes cannot be larger in both dimensions. The analysis of Fig. 2.6 could be repeated for each of these boxes.

Hence, even with fixed claims on the goods produced resulting from the distribution of productive resources between households, the two efficiency conditions yield many combinations of resource allocations and distributions of the resulting production. These constitute a subset of the possible configurations of the economy. But within this subset of efficient resource allocations there is a possibility that some of these combinations may be dominated by others, in the sense that some allocations of factors of production with their resulting patterns of goods, when distributed to the households, may put all of them on higher preference contours than other efficient combinations. A third efficiency condition is needed to ensure against dominance. We thus have three conditions for resource efficiency in an economy:

1. available factors of production are allocated so as to put the economy on its PPF;

2. gains from trade are exhausted;
3. no allocations satisfying 1 and 2 dominate others in terms of household preference.

If an economy allocates its resources so as to satisfy these three conditions it is said to be at a *Pareto optimum* and the resulting pattern of production and allocation of goods is *Pareto efficient*.

If the economy is at a Pareto optimum, then it is not possible, by any realloca-tion of resources or any redistribution of goods, to make one household better off without making another worse off.

This is the economists' notion of efficiency for an economy. When an econ-omy is at a Pareto optimum, households, given the resources at their disposal, their preferences for goods, and the trading opportunities represented by the preferences of other households and the goods that they possess, would not freely choose any other configuration of the economy. Each household is maxi-mizing its welfare, subject to these constraints.

A Pareto optimum is defined for a given set of claims to resources as repre-sented by the initial allocation to point P in Fig. 2.6. We have seen that a different starting point deriving from a different distribution of ownership of, and claims on, resources would affect household welfare. In discussing welfare, therefore, economists usually distinguish between allocative efficiency, which is achieved through Pareto efficiency, and distributional considerations. Logically, if distribution is considered just then society should prefer organiza-tions of the economy that were allocatively efficient to those which were not. However allocatively efficient organizations may be distributionally unaccept-able, if resources are concentrated on a small minority of the very rich to the exclusion of a large number of the very poor.

A flock of cooperating redshanks might exploit the available feeding area very efficiently, distributing their efforts across time and space so that the collective catch of prey items, given the constraint on foraging time, could not be increased. However if the food collected was taken to a central point and fought over, with the result that 90% of the food was consumed by the dominant males and the juveniles in the flock starved, ecologists might question the efficiency of the strategy! Society's welfare has both an allocative and a distributional aspect.

But redshanks are not social animals and our concept of efficiency has gone a long way from the notion of efficiency for the foraging redshank. The com-plexities arise in part because humans are social animals practising a division of labour. This not only introduces a distributional dimension to the assessment but it complicates the allocative rules, since efficiency must be specified for all the cooperating individuals and efficiency rules are not 'separable' in the sense that we can specify conditions for an individual which will stand for the collec-tive. This problem is complicated further by the fact that humans can trade as well as fight over the collective catch. Further complexities arise because humans are tool users and have therefore more than their labour time to allo-cate. One aspect of this problem that we have not discussed concerns the allocation of resources between consumption and savings. This is briefly dis-cussed at the end of the chapter.

A monetary economy and markets

In the simple economy so far described, the goods that households bring to the process of exchange are determined by their resources and hence their contribution to the economy's production. The exchange portrayed is barter, the direct exchange of goods. This is a convenient heuristic simplification. In a real economy very many goods and services are produced with a complex division of labour and all households, numbering millions, participate in exchange. With such a vast and complex system barter is infeasible. Complex exchange is made possible by the existence of money. Money performs three functions in an economy: it acts as a medium of exchange, a store of value and a unit of account. Households sell their resources, or the products of them, for money which, in turn, they use to acquire goods. Goods and services are bought and sold in the marketplace and command money prices. Exchange rates are expressed in ratios of money prices. A complex economy is a series of interlocking markets: for factors of production (labour markets and capital goods markets) and for consumption (goods and services). In all of these markets money is the medium of exchange and factors and goods traded command money prices. In a monetary economy with organized markets it is not necessary for a household wishing to sell bread and buy blankets to find a household wishing to sell blankets and buy bread; they sell bread for money in the bread market and buy blankets for money in the blanket market. Specialized traders operate in both markets. Thus money permits the existence of organized markets and economizes on the information needed by households engaging in trade.

A market for a good is usually portrayed as the interaction of the forces of demand and supply. The demand function gives the quantities of the good that households will wish to buy at various prices; the supply function gives the quantities of the good that suppliers will offer for sale at various prices.

Demand curves are normally negatively sloped. Households will demand more of a good as its price, relative to the prices of other goods, falls. There are two reasons for this:

- A fall in the price of a good makes the household better off. The purchasing power of its money income (derived from the sale of the services of its factors of production) will increase. This increase in purchasing power will cause the household to consume more of all goods including the one whose price has fallen.
- The good has become cheaper relative to others in the household preference function so it substitutes this good for others.

Supply curves are typically positively sloped. More will be offered for sale as price rises. The reason is that increased production requires shifting resources (factors of production) from other goods and this faces rising opportunity cost. In a monetary market economy, specialist producers own capital and hire labour services and produce for *profit*. Profit is the difference between total revenue from sale of the good or service produced and the total money cost of producing it. Profit is increased as long as the additional cost of increasing output exceeds the revenue it generates. In a competitive market, incremental cost from increasing output is termed marginal cost and the competitive producer's supply curve is her marginal cost curve.

In well-behaved markets demand and supply curves will intersect giving the equilibrium price (supply equals demand). *Demand and supply are functions of relative not absolute prices* so that changes in the prices of other goods and in consumer money incomes, which represent the rewards for supplying factors of production, will cause these curves to shift.

Figure 2.7 shows supply and demand curves in the market for a good. P^* and Q^* are the price and quantity in equilibrium at which the market 'clears'. The dotted lines show hypothesized shifts in the curves as a result of a rise of prices of goods in other markets. Demand in the market portrayed will shift to the right implying that households will wish to purchase more at every price since the good has become cheaper compared with others. On the other hand the rise in the prices of other goods will reduce the purchasing power of household incomes and demand in the portrayed market will shift to the left as households are now poorer. As depicted the former effect dominates and the new demand curve is D'.

The rise in the prices of other goods will shift resources into production in those markets. This will cause the opportunity cost of resource use in producing goods in the market portrayed in Fig. 2.7 to rise. Hence the supply curve shifts to the left indicating that less will be offered for sale at any price.

The portrayed shifts in practice are likely to be small and are (deliberately) exaggerated in the diagram.

Competition and efficiency

Efficiency in markets is ensured by *competition*. In competition, sellers are free to enter markets whenever they see opportunities for beneficial trade and to leave them should they perceive the opportunities to be exhausted. Competition drives down prices and ensures efficiency. Provided that competition prevails in all markets the resulting configuration of the economy will be

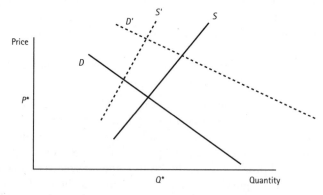

Fig. 2.7 Market demand and supply.

Pareto optimal: it is not possible to make any subset of households better off without simultaneously making another subset worse off. The situation of free competition in all markets is called *perfect competition* and a perfectly competitive economy provides the benchmark of efficiency against which actual economies are judged.

A measure of the welfare effects of competition is given by what is called *consumers' surplus*. The demand curve indicates what the households (consumers) would have been willing to pay to obtain each unit of the good. A market demand curve is a summation of the demand curves for all of the households that are in the market.

Figure 2.8 gives the demand schedule for a individual household for a hypothetical good. It shows that it would have been willing to pay the high price P_a to obtain a small quantity of the good and if P_a had been the market price Q_a would have been purchased. P_b is a lower price at which a larger quantity Q_b would have been purchased. As the price falls so the quantity that households want to buy rises. P_c is the low competitive price and Q_c is the quantity that the household will buy in the competitive market. In an organized market one price prevails. The demand curve shows that consumers would have purchased some of their supplies at higher prices, i.e. they receive some of the commodity at less than they would have been willing to pay for it. Thus, if P_c is the market price the household receives quantity Q_a at that price even though it would have been willing to pay the higher price P_a for it. The difference between what it is willing to pay and the price actually paid is called the consumers' surplus and is the area under the demand curve above the price line. It represents purchasing power available for other goods. Obviously the lower the price in the market, the greater the consumers' surplus. Thus consumer welfare is increased in organized markets by the opportunity to buy goods at a fixed price. Competition forces that price to the lowest level compatible with the supply of the good; it therefore maximizes the consumers' surplus.

Investment and savings

One aspect of the household's resource allocation problem briefly mentioned earlier concerned the allocation of time and other factors of production

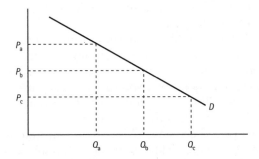

Fig. 2.8 The consumers' surplus.

between production for current consumption and the creation of capital goods. For the isolated household, production of capital goods reduces the resources available for current consumption but increases those available for consumption in the future. This activity can be viewed from either perspective. The act of setting resources aside from production for current consumption is termed saving, and the production of additional resources for future production and hence consumption is termed investment. Since these are the same activity viewed from different perspectives, savings and investment are necessarily equal.

This equality between saving and investment for the isolated household must also hold for society as a collection of households. However the complexity of the social organism necessarily complicates and confuses the issue in many respects.

1. In a monetary economy saving is achieved by setting money aside from current expenditure. This can be done because (heroically leaving aside the problem of inflation) money can be stored and used to acquire goods in future periods. This is the second role of money listed above: money as a store of value.
2. The division of labour means that the acts of savings and investment are not necessarily carried out by the same economic units and savings and investment intentions must be reconciled through markets. Specialized institutions exist to marshall savings (financial institutions, e.g. banks) and to transmit them to those who carry out investment.
3. If the markets fail to reconcile savings and investment intentions then resources are not utilized efficiently. Abstentions from consumption, savings, if not matched by investment, are wasted and society has forgone consumption opportunities for no return.
4. The range of types of investment is much larger. Investment may be made in raising the skills and capacities of labour (education and training) and in acquiring knowledge of new techniques and products (research and development). Investment is any activity that pushes out society's PPF, increasing the quantity, quality and range of goods and services available to it.

The field of investment and savings is a specialized and substantial area of economics that is of only marginal relevance to the issues addressed in this book. I confine myself here to no more than some basic concepts relating to the notion of efficiency of resource use.

Even for the isolated household the issue of how much of current resources to devote to investment is a relevant one. Efficiency requires consideration of two aspects of the process: the productivity of investment, and time preference.

Forgoing current consumption and devoting resources to investment increases future consumption possibilities. The productivity of investment measures the relationship between the sacrifice and the future gain. Figure 2.9 illustrates the idea for a simple two-period decision. C_t is current consumption and C_{t+1} is future consumption. If the household devotes all of its resources to production for current consumption it will consume C_t^*. Devoting all resources to future

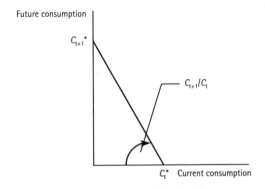

Fig. 2.9 The productivity of investment.

consumption will give it C_{t+1}^* in the future. C_{t+1}^* is larger than C_t. The line joining these two points is the households *inter-temporal consumption possibility frontier*. The slope of the line C_{t+1}^*/C_t^* measures the productivity of the investment; it says how much future consumption may be obtained for a sacrifice of a unit of current consumption. It is usual to express the productivity of investment as its rate of return. The rate of return on the investment is the productivity minus 1 expressed as a percentage: $[(C_{t+1}/C_t)-1]^* \times 100\%$. Thus if the saving of 100 units of current consumption yields 110 units of future consumption the rate of return is $[(110/100) - 1] \times 100 = 10\%$.

As well as the rate of return on possible investment, the household needs to take account of its preferences for present and future consumption. Figure 2.10 shows contours of an *inter-temporal preference function* for a household. All points on contour 2 are preferred to points on contour 1 since it offers more of both current and future consumption. The slope of a contour shows the household's trade-off between current and future consumption, i.e. the quantity of additional future consumption that it requires in return for a sacrifice of a unit of current consumption if it is to retain its level of welfare.

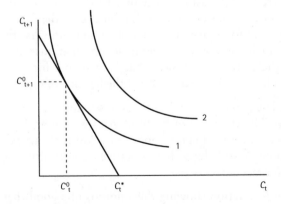

Fig. 2.10 Inter-temporal efficiency in resource allocation.

The curve steepens towards the left implying limits to the willingness of the household to forgo current consumption. The level of saving at any point on the curve is the amount of current consumption forgone. The slope of the curve at that point is the productivity of saving that the household requires if it is to undertake that amount of saving. By analogy with investment, that productivity can be expressed as a rate of return on saving. This is known as the household's *rate of time preference*. Resource efficiency is achieved where the rate of return on the investment equals the rate of return that the household requires on its saving: its rate of time preference. This is given by a level of current consumption C_t^0 and a level of saving $C_t^* - C_t^0$, where the contour of the inter-temporal preference function is tangential to the investment function.

In a monetary economy where households save and producers invest, the households determine their savings decisions by the rate of return, also known as the rate of interest, that financial institutions offer them on their savings. The financial institutions then transmit these funds to the investment market and investors seek rates of return that cover the cost of acquiring the finance for that investment. The rate of return is thus the price of savings and the market clearing price equates savings with investment.

Glossary

Allocative efficiency A situation for an economy where it is on its production possibility frontier and there are no possible gains from trade.

Capital A factor of production. Tools and machines etc. used for the production of goods and services that have been produced, using factors of production, in previous periods. The opportunity cost of capital is consumption since some of the resources available for the production of consumption goods have to be diverted to the production of capital.

Capital good A specific good, typically a tool or machine that is a component of capital.

Consumers' surplus A measure of the benefit that consumers derive from being able to trade at a constant price. It is indicated by the difference between what is actually paid for a given quantity of a good and what the household would have been willing to pay in order to acquire that quantity. It is thus an expression of the purchasing power that is available for other purposes. Consumers' surplus is thus an expression of the benefits of organized markets.

Consumption The goods and services that households acquire and use.

Contract curve The subset of efficient choices within an Edgeworth box. Between points on a contract curve opportunity cost is always positive. Points off the contract curve will always have negative opportunity costs with some points on the contract curve.

Demand curve A function showing the amounts of a good that buyers will wish to buy at various prices. Demand curves exist for individual households and firms and they may be aggregated to give total or market demand curves. Demand curves are usually negatively sloped indicating that consumers will buy more at lower prices.

Division of labour Specialization of individuals in specific production tasks. The division of labour increases productivity and hence the quantity of goods and services that can be produced with available resources. The cost of the division of labour is a loss of self-sufficiency. Households are then dependent on trade to meet some or all of their needs.

Edgeworth box A graphic technique for portraying choices and showing the difference between efficient and inefficient choices. The dimensions of the box represent the set of available choices.

Efficient technique One that for a given output of a specific good or service uses less of at least one factor of production than other available techniques. From the principle of scarcity only efficient techniques will be used.

Factors of production A classification of the resources that are used in the production of goods and services. These are conventionally categorized into land, labour and capital.

Firm The basic production unit in an economy.

Gains from trade A situation where households could mutually gain from exchanging goods. The exhaustion of gains from trade is a necessary condition for obtaining a Pareto optimum.

Good A tangible item of consumption (a chair, a loaf of bread).

Household preference function The function showing household preferences between various combinations of goods and services. From the principle of scarcity households will wish to reach the highest attainable contour of their preference function.

Household The basic consumption unit in an economy.

Indifference curve A contour of a household's preference function showing the combinations of good between which the household is indifferent and which therefore yield it the same level of welfare.

Inefficient technique One that for a given output of a specific good or service uses more or no less of all factors of production than other available techniques. From the principle of scarcity inefficient techniques will not be used in production. Techniques that were previously efficient may become inefficient as a result of technical progress.

Inter-temporal consumption possibility frontier The outer boundary of inter-temporal consumption possibilities. It shows the trade-off between future and current consumption achievable as a result of investment.

Investment The creation of capital. Investment for an economy is constrained by the volume of savings. In a subsistence economy investment and savings are one and the same thing. In a complex monetary economy investment and savings are carried out by different units. Investment is done by firms or the state. Savings are made by households.

Isoquant A contour of a neoclassical production function. All the available combinations of factors of production for producing a given quantity of a good or service.

Labour A factor of production. Human time devoted to the production of goods and services, i.e. work.

Land A factor of production. It comprises the physical space necessary for any economic activity. It is often defined to cover the gifts of nature which are also used in production.

Market A device for the exchange of goods and services or for factors of production. In a market trade takes place.

Neoclassical production function A production function in which factors of production are assumed to be continuously substitutable so that the number of available techniques of production is indefinite.

Opportunity cost The value of a choice in terms of the alternative opportunities forgone in making that choice.

Pareto efficiency An arrangement of an economy satisfying the conditions for a Pareto optimum.

1. Available factors of production are allocated so as to put the economy on its production possibility frontier;
2. Gains from trade are exhausted;
3. No allocations satisfying 1 and 2 dominate others in terms of household preference.

Pareto improvement A change in the allocation of resources in an economy that makes some households better off and none worse off.

Pareto optimum A situation for an economy where it is not possible, by any change in the combination of goods produced or by any change in the distribution of goods between households, to make some households better off (in the sense of putting them on a higher indifference curve) without simultaneously making others worse off. A Pareto optimum exists if three necessary conditions are satisfied.

Perfect competition A situation where all potential sellers (and buyers) are free to enter and leave markets if they perceive opportunities for beneficial trade. Price is bid down to the lowest level compatible with continuing production. If all markets are perfectly competitive then the economy will obtain a Pareto optimum. Under perfect competition consumers' surpluses are at a maximum and perfect competition is thus treated as the yardstick against which to measure allocative efficiency.

Potential Pareto improvement A possible change in an economy's resource allocation so that if it is carried out the gainers could compensate the losers and still be better off. If the gainers do compensate the losers when the change is made then the change constitutes a Pareto improvement.

Production function The set of all the efficient techniques available at a given time for the production of a specific good or service.

Production possibility frontier The outer bound of production possibilities for an economy. Between points on the frontier opportunity cost is always positive. Opportunity cost will be negative between points within the frontier and areas on the frontier. On the frontier it is not possible to produce more of some goods without producing less of others. Being on the frontier is a necessary condition for Paretian efficiency.

Productivity of investment The amount of future consumption achievable per unit of investment in the current period.

Productivity of savings A measure of the additional future consumption that a household can achieve as a result of forgoing current consumption (i.e. saving).

Productivity Output per unit of factor input.

Rate of return on investment 1 minus the productivity of investment expressed as a percentage. It is formally the rate of discount that will make the present value of the anticipated profits from an investment equal to the capital cost. The notion is explained at length in Chapter 9. The rate of return is the opportunity cost of an investment and firms will only invest if the rate of return is anticipated to exceed the rate of interest that they must pay on borrowing to finance the investment.

Rate of time preference The rate of return that a household requires if it is to practise savings. It is equal to 1 minus the productivity of savings expressed as a percentage. To attract savings from households, financial institutions must offer a rate of interest equal to or above the time preference rate.

Savings Resources not consumed but made available for adding to the stock of capital. In a capitalist society households set aside parts of their incomes for savings and receive in return for them payment in the form of interest credited. Specialist financial institutions (e.g. banks) exist to collect savings and transmit them as loans to firms that use them to create capital.

Scarcity The principle of scarcity says that resources are scarce relative to the uses to which they can be put. Because of this resources have opportunity costs. Scarcity is the ultimate source of value in society since any choice involves sacrifice of alternatives.

Service An intangible item of consumption (haircuts, bus journeys).

Supply curve A function showing the quantities of a good or service that suppliers will offer for sale as a function of price. Supply curves are normally assumed to be positively sloped indicating that at higher prices greater quantities will be offered for sale.

Technical progress An increase in the available techniques of production.

Technique of production A formula or recipe for producing a specific quantity of a good or service in terms of the quantities of factors of production used.

Trade The exchange of goods or services.

Chapter 3

Market failure and the environment

Neoclassical environmental economists argue that environmental problems arise from market failure and define the optimum state of the environment as that which would hold were the sources of market failure corrected. From this diagnosis of the nature of the problem comes a prescription of treatment, policy instruments, such as pollution taxes and marketable permits, which correct the market failure and bring about the optimum state of the environment. The argument is developed particularly for problems of pollution, but in modified form it is applied to other environmental problems that are not classifiable as pollution. This doctrine is the subject of Part 2 where it is explained and critically examined. The starting point must be to explain the notion of market failure. That is the function of this chapter.

This chapter assumes an understanding of the notion of efficiency in economics and the neoclassical doctrine that efficient resource allocation is brought about by the force of competition in competitive markets. According to this doctrine the purpose of markets is to achieve efficiency, i.e. to afford households the opportunities to enhance their welfare through exploiting trading opportunities and to ensure that goods and services are produced efficiently and exchange at the lowest prices consonant with the requirements of production and trade. For those without previous experience of economics these ideas are explained in Chapter 2.

Departures from allocative efficiency are termed market failure and arise when the conditions for competitive markets to achieve efficiency do not hold. There are three main causes of market failure:

1. monopoly;
2. the public goods problem;
3. externalities.

I examine these in turn.

Monopoly

Monopoly occurs where there is only one supplier of a commodity and that supplier is able to prevent others from entering the market and competing with it. Monopolists possess market power in that, by their actions, they are able to influence the price prevailing in the market. In other words, monopoly exists when a seller of a good or service is not forced, through competitive pressures, to accept the market price. Hence perfect competition is a state where no one possesses market power and the requirement for efficiency, for a Pareto optimum, is alternatively viewed as a lack of market power. Monopolists use their power to raise the price and thus to capture some of the consumers' surplus as what is termed monopoly rent. This is shown in Fig. 3.1. Although monopolists are able to capture some of the consumers' surplus by raising the price above the competitive level, they cannot capture it all unless they are able to charge a different price to each consumer. Under monopoly some of the surplus is lost in the process. The lost surplus is the triangle in the diagram. The monopolist's gain is less than the consumer's loss. The monopolist captures part of that surplus as monopoly rent but part of the surplus is lost as an inefficiency because the conditions for efficient resource allocation are not met. What has happened is that the monopolist can only raise the price at the expense of the quantity sold and the resulting output is less than the efficient (competitive) output.

This problem of market failure can be viewed from the other end. In the presence of monopoly it is said that there exists a potential Pareto improvement. If a policy were adopted that eliminated the monopolist's power and reduced the market price to the competitive price, the gainers from this policy, the consumers, could afford to compensate the loser, the monopolist, and still be better off. The difference between the minimum compensation that monopolists would need to receive were they not to lose out from this policy and the gains to the consumer is termed the social surplus. In Fig. 3.1 the social surplus is again the triangle of lost surplus, i.e. the difference between the monopoly rent and the consumer surplus lost in the course of acquiring it.

All forms of market failure hold out the prospect of a potential Pareto improvement should the failure be corrected. However correction of the failure does not necessarily result in an actual Pareto improvement. An actual Pareto improvement occurs if the change results in some parties being better

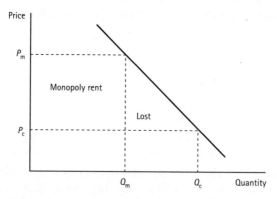

Fig. 3.1 The monopoly problem. P_m, Q_m, monopoly price, quantity.
P_c, Q_c, competitive price, quantity.

off and none worse off. To convert a potential Pareto improvement into an actual improvement it is necessary that the sufferers from the change actually be compensated. In the case of the monopoly, the potential Pareto improvement would be converted into an actual efficiency gain if monopolists were compensated for their lost profit. Thus it would not be sufficient, say, for the Government to introduce price controls to force the monopolist to charge the competitive price. It would actually have to tax the consumers and use the revenue to compensate the monopolist for the lost profit. If it did that then there would be an actual efficiency gain from eliminating the monopoly.

Monopoly is not a source of market failure that is cited in the analysis of environmental problems, although it does play some role. The major forms of market failure relevant to the environment are the public goods problem and externalities.

Public goods

The conventional analysis of the problem of allocative efficiency as given in Chapter 2 is based on the unstated assumption that all goods are private. Markets only function efficiently for the trade in private goods. A private good has two essential characteristics: it is excludable and it is rival.

A good is excludable if the seller can exclude non-buyers from its consumption. In essence, in a market transaction the seller transfers property rights in the form of use rights to the buyer. The buyer then can consume those goods for which she has acquired property rights, give them away or do with them what she will.

If the seller cannot transfer these exclusive use rights, then the rational consumer will not trade. Instead she will consume the goods without purchasing them. This is known as *free riding*. It is rational since, if the buyer can consume the goods without purchase, then the money that would have been spent is available for something else. Thus the principle of scarcity predicates free riding where goods are non-excludable. In the presence of free riding the seller cannot exact a price for his product. Free riding causes markets to fail and goods that are non-excludable will be under-supplied or not supplied at all by the free market.

The notion is probably best illustrated by some examples.

- The provision of a nuclear deterrent for a country is a non-excludable good. Everybody living in the territory is protected by the deterrent whether they wish to be or not. It is thus not possible to sell nuclear protection to individual citizens or households since those who do not buy are equally protected with those who do. In consequence people would free ride, claiming that they did not want protection in order to avoid payment. Of course citizens who genuinely reject the theory of nuclear deterrence, believing that the possession of nuclear weapons increases rather than reduces the threat of nuclear attack, cannot exclude themselves either. For them nuclear deterrence is a public bad, reducing their security and hence their welfare.

- A less controversial example would be the provision of lighthouses to warn shipping of dangerous rocks and to assist in navigation. The services of a lighthouse cannot be sold to the captains or owners of shipping since those who do not pay will benefit equally with those who do. All shipping navigating the waters where the lighthouse is located benefits from its services whether or not it contributes to its cost.

Non-excludability is a primary reason for collective or public provision of goods and services. Even so, free riding remains a problem since the public will not reveal their true demand for these goods if they perceive that the cost of provision through taxation will be related to demand. Thus with non-excludability market failure takes the form of the under-provision of these goods in comparison with other excludable goods. The public can, and does, simultaneously complain about the quantity and quality of non-excludable public goods and services and resists increases in taxation which would be required to improve the provision. Economists have devised ingenious taxes that overcome the problem of free riding but these devices are simply theoretical solutions to the problem. Their complexity makes them impracticable in real-life situations.

A good is rival if, in transferring the good to one purchaser, the quantity the seller has available for sale is reduced. If this is not the case and the good is non-rival, then the opportunity cost of supplying an additional buyer is zero since no resources are involved in the provision. If there are many producers of a non-rival good, competition would in consequence force the price of the good towards zero and, if the provision of the good in any quantity has costs for the seller, i.e. its production uses resources, she will not be able to profitably supply it under these circumstances.

If non-rival goods are excludable they will be provided through markets and the seller will attempt to average her costs over the quantity supplied. But this process will exclude some buyers who would be willing to pay less than the price asked but more than the (zero) cost of providing them. This is the source of market failure with non-rival goods.

Many excludable goods are non-rival over ranges of consumption. Thus the services, in the form of observation of flora and fauna, provided by a nature reserve (up to the point where user pressure starts to damage the resource) are non-rival. Users can be excluded from entry – nature reserves are excludable goods – but they are non-rival. Hence private provision would not be optimal since some consumers who value the service at less than the entry charge are excluded.

Another example, familiar to transport economists, concerns the charging of tolls for trunk roads and bridges. Provided the road is not congested, allowing an extra vehicle to use it imposes trivial costs on the operator. The toll therefore will exclude users who value its use at less than the toll charged. But the toll is designed to cover the cost of provision over the anticipated volume of user traffic. Maximum consumer benefit is achieved by providing the road free but then the market would not provide the facility. The problem with excludable but non-rival goods is that of what are termed sunk costs. Once resources are committed to building the road they are said to be sunk. The resources cannot be taken back and no further resources (I ignore for the present the issue of maintenance) are required to be committed.

These examples of excludable but non-rival goods only possess those characteristics over a range of consumption levels. Once traffic volumes have risen to the capacity of the road, congestion sets in and an additional user imposes real resource costs (in the form of extra journey time and vehicle operating costs) on all users. Equally, over-use of the nature reserve causes a fall in the quality of the experience for all users and may additionally damage its flora and fauna. Most, if not all, excludable but non-rival goods are non-rival only over a range of consumption. Beyond those levels an additional road has to be provided or an alternative reserve and this has real costs.

Nuclear defence and the lighthouse service, the examples that I used for non-excludability, are also non-rival goods. The cost of providing nuclear deterrence to another citizen, or lighthouse facilities to another ship, are zero. Furthermore they are non-rival over all ranges of consumption. A dangerous rock does not need two lighthouses, however heavy the shipping, and nuclear deterrence is related to an area of land and not to the number of people living there. These goods are members of a class of what can be termed *pure public goods*. Where goods are either non-rival or non-excludable but not both we have what may be termed *partial public goods*.

The class of pure public goods contains a substantial number of members. As we have seen there are also many partial public goods that are excludable but non-rival within a range of output. There are few examples of the opposite case: goods that are rival but non-excludable. The classic and famous case concerns bees (*The Fable of the Bees*, de Mandeville, 1924). A modern version of the famous fable may be in order.

Colonial bees, honey-bees (*Apis*) and bumble-bees (*Bombus*), are important pollinators and their services are required in large numbers in, for instance, fruit orchards when the trees flower. The commercial supply of honey-bee hives to place in orchards at the crucial time is dependent on the market for honey and is subject to a degree of uncertainty. In intensive arable farming areas *Bombus* bees are thinly scattered and tend to be confined to areas of settlement and woodland. They are, however, easy to rear. Colonies die out in the autumn and the fertilized young queens over-winter. They are easily caught in the spring and will readily start their colonies if given the right environment. The colonies are small, the inhabitants docile, and much more transportable than honey-bees. Since bumble-bee colonies do not over-winter they do not produce an exploitable crop of honey but they none the less perform the useful service of pollination.

A specialist market in bumble-bee colonies does not arise because colonies have the characteristic of partial public goods. The services of the worker bees are clearly rival: it takes time for a worker to visit a flower and there is obviously a limit on the amount of pollination that a colony can achieve in a day. The opportunity cost of pollinating one tree is another tree not pollinated. But the services of the bees are non-excludable. A supplier has no means of programming the workers to visit orchard trees rather than some altogether different plant species within their foraging area. In other words a putative seller of bee services could not guarantee the bees' services to a buyer. He is unable to sell the property rights of pollination to the buyer since he does not himself own them. Hence because of the public goods problem there is no market in which you can buy or rent the services of bumble-bees. For the same reason there is no market in the services of honey-bees either but because they produce a honey crop there are bee-keepers who are willing to place their bees in orchards and other sources of pollen and nectar without charge.

The fable of the bees is probably extensible to other attempts at marketing wildlife either as spectacle or for exploitation. The seller of wildlife as either spectacle or 'sport' cannot guarantee that the buyer will see or (still less) kill a particular species of animal on a particular day. This public goods problem shapes the way in which wildlife is marketed and probably limits the size of the market. Either the uncertainty is made a central part of the package ('the chance of seeing a golden eagle'; those who want certainty will go to a zoo), or the diversity of possible species is emphasized (and the probability of encountering at least some of them is then sufficiently high), or wildlife experience is sold with other goods (trekking, comfortable hotels, education). With commercial slaughter, grouse and pheasant shooting for instance, the non-excludability problem would be severe because the customers tend to have a high valuation of their time. It is overcome by the use of intensive land management, which raises the population of the prey species to levels where success can be guaranteed.

If a good is non-excludable but rival then free riding means that a price cannot be charged even though the optimum price, that which delivers the quantity that consumers wish to consume, is positive. With goods that are non-rival but excludable the optimum price is zero but private provision will ensure that the price is positive and the quantity consumed below the optimum. Thus with non-excludable goods free provision is necessary but not optimal. With non-rival goods free provision is optimal but not necessary. With pure public goods, since they are both non-excludable and non-rival free provision is both necessary and optimal.

The public goods problem is part of the economist's analysis of environmental problem and is examined further in subsequent chapters.

Externalities

The argument has proceeded so far on the implicit assumption that households are independent actors in the economy. Their welfare is dependent on the goods and services that they themselves consume and their command of those goods and services is dependent on their income, which is derived from the sale or use of the factors of production they possess. In this simple model the only way that a household can affect the welfare of others is through markets and indeed it is solely through markets that they interact. Where that is not the case, where the welfare of economic units (households or firms) is affected by the economic activities of other units in ways other than through markets, we have externalities and competitive markets will not ensure Pareto efficiency.

Externalities can be positive, i.e. one unit's activities increase the welfare of other units, or negative. If an increase in the consumption of a particular good by one household raised the welfare of another then we would have a positive externality and in choosing the quantity of the good to consume the first household would be conferring a social benefit. If, on the other hand, an increased consumption of this good by a household lowered the welfare of others, we would have a negative externality and the household's decisions on the consumption of the good would carry a social cost. Box 3.1 gives an algebraic statement of externalities.

Box 3.1 A formal statement of externalities

Formally the model for a two-household (A,B) economy is:

$$W^A = f(X_1^A \ldots X_n^A)$$
$$W^B = f(X_1^B \ldots X_n^B)$$
$$X^A = f(Y^A, P_1 \ldots P_n)$$
$$X^B = f(Y^B, P_1 \ldots P_n)$$

where W is the household's welfare; $X = (X1 \ldots Xn)$ are goods and services that the households consume; $P1 \ldots Pn$ are the prices of these goods and services; and Y is household income.

If household A's welfare was additionally dependent on some good or service under the control of B:

$$W^A = f(X_1^A \ldots X_n^A, X_i^B)$$

then we would have an externality. In determining its welfare, household B, through its consumption of good X_i would affect A's welfare.

Externalities are classified by the sign of the partial derivative $\delta W^A / \delta X_i^B$. If $\delta W^A / \delta X_i^B > 0$ we have a positive externality and B's decision on the quantity of X_i^B to consume confers a social benefit on A. If $\delta W^A / \delta X_i^B < 0$ we have a negative externality and B's decision on the quantity of X_i^B to consume carries a social cost.

Externalities can be between households as in our example, e.g. playing loud music disturbs neighbours, or between producers, e.g. discharges of polluting substances into a water course adversely affects the activities of a downstream producer who requires clean water, or between consumers and producers, e.g. black smoke from a factory dirties washing hung out to dry.

In the presence of externalities, positive or negative, competitive markets do not yield efficiency and gains from trade are not exhausted. This failure to eliminate gains from trade is why externalities are described as a form of market failure.

Environmental problems are typically presented and analysed by economists as externalities and the analysis of externalities is explored in greater depth in the next chapter on the economic theory of pollution.

Summary

- Markets are said to fail when the conditions necessary for perfectly competitive markets are not satisfied.
- In the presence of market failure resources are not allocated according to the requirements of Pareto efficiency. Hence there are potential Pareto improvements from the correction of market failure.
- Neoclassical environmental economics views environmental problems as examples of market failures and defines optimum states of the environment as those that would prevail if the market failures were corrected.
- Market failures are classified into three main types: monopoly, the public goods problem and externalities.
- The monopolist seeks to capture some of consumers' surplus for himself but in so doing causes efficiency losses. Monopolized goods will be under-provided in comparison with what would occur in competitive markets.
- For markets to operate, goods must be both rival and excludable. The public goods problem occurs when one or other of these conditions, or both of them, are not satisfied. Where neither is satisfied we have pure public goods; otherwise, where only one is satisfied we have partial public goods.
- A good is rival if there is a positive opportunity cost to the seller in providing it to a buyer. It is excludable if the seller can exclude non-buyers from consumption or, in other words, can confer exclusive property rights to the goods that she sells.
- Non-rival goods will be under-provided in markets; non-excludable goods will not be provided at all since consumers will free ride.
- Externalities exist where the decisions of economic units (households, firms) affect the welfare of others in ways other than through markets. Depending on whether those affected benefit or suffer, externalities are classified as either positive or negative and the decisions that cause them are giving rise to either social benefits or social costs.
- In the presence of externalities markets fail because gains from trade are not exhausted.

Part 2

The choice of instruments for environmental policy

Chapter 4

The economic theory of pollution

Pollution is a classic case of an externality and the economic theory of pollution is built around this. For the simple theory we assume two parties: a single polluter and a single sufferer. The analysis is best explained with a simple diagram (Fig. 4.1). The polluter, designated A in the diagram, is a factory which in the process of production produces pollution that reduces the welfare of the sufferer, designated B. It is not necessary for this analysis to specify what the pollution is. The factory may be producing smoke, an offensive smell, or loud noise, which upsets a nearby resident. Alternatively the receiving medium may be water rather than air and the factory may be discharging some noxious substance into a river that is used by the sufferer who is located downstream of the factory. The analysis is general and is intended to cover all of these cases.

In Fig. 4.1 pollution is measured horizontally, increasing from left to right. Producing pollution is beneficial to the polluter since it is an incidental by-product of production and the production yields revenue to him. The polluter's benefit from producing pollution is indicated by the willingness to pay curve (WTP$_A$ in Fig. 4.1). WTP$_A$ is in effect the demand curve of the polluter for pollution. It indicates what he would be willing to pay for the right to emit various levels of pollution. In the absence of any constraint on his activities the right to pollute by the polluter is free. In these circumstances he will pollute until the benefit from further pollution is exhausted. Line OC of pollution achieves this since, at C, WTP$_A$ is zero.

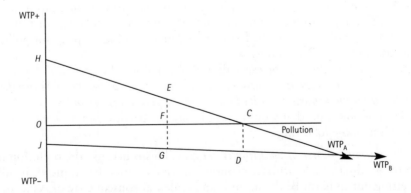

Fig. 4.1 Pollution as an externality.

But if pollution yields a benefit to the polluter, it produces the reverse for the sufferer. WTP_B is the sufferer's evaluation of the pollution and represents her demand curve for pollution. Since pollution is bad, not good, for B her WTP curve is wholly negative. This might be interpreted to mean either that B is willing to pay *not* to have pollution or that, if she is to suffer pollution, then she requires the payment of compensation if her welfare is *not* to suffer. The issue of interpretation is discussed anon. WTP_B is drawn with a negative slope, which means that the sufferer's assessment of the damage that she suffers from further pollution increases with the level of pollution. This assumption is not necessary and whether it is realistic will depend on the nature of the pollution. WTP_B could be horizontal indicating each unit of pollution inflicts the same amount of damage to the sufferer. It is possible even that WTP_B could have a positive slope after some point indicating saturation for the sufferer with further pollution adding less and less to the damage she suffers. These differences will not affect the analysis provided that WTP_B never becomes positive.

If the polluter, A, decides how much pollution to produce then we say that A possesses the *property rights* in the pollution (i.e. that A has the right to pollute even though B suffers from the pollution). In the absence of specific policies to control pollution property rights *de facto* reside with the polluter.

We have seen that A will prefer to produce *OC* of pollution since at *C* his WTP is zero and his benefit from polluting is at a maximum. At any point to the left of *C* the polluter would gain from increasing pollution. Any increase in pollution beyond *OC* will however reduce his welfare since his WTP is negative. By producing *OC* of pollution, A receives a maximum consumer surplus equal to the triangle *OCH*.

With pollution *OC*, B suffers a loss of welfare measured by the negative consumer surplus of *OJDC*. At *C* household B would be willing to pay an amount *CD* to reduce pollution. This is greater than A's zero WTP for pollution at that point. Hence, *in this situation there are gains from trade*. If B pays A any amount less than or equal to *CD* to reduce the pollution that B suffers then both parties will be better off. Since gains from trade are not exhausted the situation where A produces *OC* of pollution is not Pareto optimal.

Gains from trade are exhausted if A's pollution is reduced to *OF* where *EF* = *FG*. At *F*, the maximum that B is willing to pay for further reductions in pollution, *FG*, is equal to the minimum compensation that A requires for further reductions in pollution. To the left of *F*, A demands a higher price for pollution reduction than B is willing to pay.

The level of pollution corresponding to *F* is termed the *optimum degree of pollution* , i.e. *the level of pollution at which the net social benefits (A's benefits from creating the pollution less B's losses from the pollution) are maximized*. It is the level of pollution that would prevail were the parties able to freely trade. About it note the following:

- *The optimum degree of pollution is not zero*. This is because the pollution is a by-product of a legitimate economic activity which has value. A is not setting out to harm B: the harm is an incidental consequence of A's legitimate activity.

- *If we move from any level of pollution to the optimum there is a potential efficiency gain,* what we have called a potential Pareto improvement. The gainer from the reduction of the pollution (B) can compensate the loser (A) and still be better off. This means that there is a social surplus in moving from *C* to *F*. That social surplus is measured by the difference between the maximum that B is willing to pay to move to *F*, the area *FCDG* (its gain in consumer surplus), and *ECF*, the minimum compensation that A would require if he is to reduce his pollution by *FC* (his lost consumer surplus). If trade takes place then that potential Pareto improvement (efficiency gain) is realized. B *must* compensate A for the losses that he bears.
- *This economic notion of an optimum has no necessary relationship to any biological notion.* The situation at the optimum may be wholly unsustainable in the sense that the pollution is in excess of the assimilative capacity of the receiving medium, however defined. Furthermore, for the economic optimum to lie within some biological maximum, it is neither necessary nor sufficient that B wishes to keep the pollution within that limit and knows what that limit is. This is because there are two parties to the trade, A and B.

A movement from *C* to *F* will realize a social surplus but it is not clear from Fig. 4.1 which party will receive that surplus. This will depend on the terms of trade, which in turn depends, in this simple two-person game, on the bargaining skills of the two parties. A will try to extract the maximum that B is willing to pay for reducing the pollution and thus to capture the entire social surplus; B, on the other hand, will try to pay the minimum that A will accept. Between these limits the outcome is indeterminate. Figure 2.6 in Chapter 2 shows that there is a range of terms of trade that would allow the parties to move to the contract curve, ranging at one extreme to terms which would move them from *P* to *Q* and at the other from *P* to *R*. Within these limits the choice of trading terms is a zero sum game, one party gaining at the expense of the other. Engaging in trade is a positive sum game, however, since both parties are made better off thereby. The present problem of eliminating the externality is exactly analogous since the parties are trading attributes of pollution: its ability to increase the polluter's welfare and to pollute the sufferer's environment.

Optimum pollution and property rights

One obvious objection to the model developed is its assumption that the polluter has the property rights, that he has the right to pollute. Many people would deny the justice of this assumption and argue instead that the sufferer, B, should have the right not to suffer from pollution. We can consider this possibility using the same diagram (Fig. 4.1) but giving B the right to decide how much pollution A can produce.

Obviously B's preference is for A to produce no pollution since in that case she suffers no welfare loss as a result of A's activity. Thus B would choose *O* as the preferred pollution level. Note, however, that at *O* there are gains from trade. A would be willing to pay up to a maximum of *OH* for permission to pollute. Since this exceeds *JO*, the minimum compensation that B would require were she to suffer pollution, trade is to their mutual advantage. If trade took place the level of pollution would increase. Gains from trade would be exhausted once pollution had expanded to *OF* since at that point the minimum B requires as compensation for permitting additional pollution is equal to the maximum that A is willing to pay for permission to increase pollution. This is of course exactly the same equilibrium as before. Thus *the optimum degree of pollution is independent of the allotment of property rights*, being determined only by the demand curves (WTP functions) of the parties to the dispute. What the allotment of property rights has done is to determine the direction of payments: *the party with the property rights receives the payment.* B pays A when A has property rights; A pays B when B has them.

This aside, the situation is symmetric. There is a potential Pareto improvement or efficiency gain from moving to the optimum. A's benefit, *OFEH*, is greater than B's loss, *JGFO*, and the social surplus is the difference between these two magnitudes. The social surplus is a different size from the case when A had the property rights but this is because the argument starts from a different point and as I have drawn it (there is no necessity for this) the distance *OF* is greater than *FC*, i.e. we are further from the optimum at *O* than at *C*. As before, how the social surplus is divided up between the parties depends on their relative bargaining skills. B must be compensated by receiving at least her valuation of the pollution, her minimum acceptable compensation (*JGFO*) but will attempt to extract the maximum that A is willing to pay (*OFEH*).

The other change that has taken place by giving the property rights to B is that we now have a positive externality (B's decision to permit pollution makes A better off) whereas we previously had a negative externality (A's decision made B worse off). In consequence if B is the decision-maker she will freely choose a suboptimal level of pollution and in permitting pollution to take place her decision carries a social benefit. If A is the decision-maker, pollution will be above the optimum and his decision imposes social costs. *Thus the classification of externalities is dependent on the allotment of property rights.*

Pigovian taxes

I have so far analysed pollution as a form of market failure. There are unexploited gains from trade that would be eliminated if the parties to the externality were able to trade. If pollution is not optimal (in terms of Fig. 4.2, the economy is stuck at *C* rather than *E*) then this must be because there are obstacles to trade. These obstacles are usually termed transactions costs. There are a number of basic categories:

- Property rights may not be agreed. A can believe that he has the right to pollute and B that she has the right not to suffer pollution; hence each believes that they are entitled to compensation from the other.
- There may be conflict over the social surplus that prevents agreement. We have seen that the maximum the party without the property right is willing

to pay to alter the pollution exceeds the minimum compensation that the party with the property right requires if it is not to be worse off. The difference between these two magnitudes is the social surplus or the gains from trade. How this will be divided between the parties will depend on their bargaining skills. Although the reduction in pollution is a positive sum game, i.e. both parties can be made better off, conflict may prevent an agreement.

- Contracts may be unenforceable and the parties may in consequence fear opportunistic behaviour. Thus B may think that if they do reach an agreement that divides up the gain, A will subsequently threaten to increase pollution unless he is paid extra. A may fear that B will not pay what is agreed.
- Negotiation costs may exceed the benefits. This can occur where there are many sufferers (not just one as we have so far assumed). As an example, consider an airport where residents living under the flight path have their sleep disturbed by night flying. The total willingness to pay of the sufferers may exceed the profits received by the airport authority from night flying, so that gains from trade exist. However there may be thousands of sufferers, each of whom is willing to pay a few pounds to stop night flying and the costs of organizing the negotiations and collecting the payments may be prohibitive.
- Free riding. The benefit of stopping night flying is non-excludable and any resident could reason that if she declares her WTP to be zero then she will not have to contribute to the payments to the airport authority but will still benefit from any agreement reached. Thus free riding will make the apparent collective WTP less than its true value so that the benefit is not realized in whole or part.

Where transactions costs prevent the elimination of the externality a control authority can achieve the efficiency gains by the use of a Pigovian tax (named after A.C. Pigou, a Cambridge economist who first formulated the idea; see Pigou, 1920, ch. vi). A Pigovian tax is a tax levied on the decision-maker, the party with the property rights, at a fixed rate equal to the WTP at the optimum ($EF = FG$). To illustrate the concept let us return to the original assumption that the polluter, A, has the property rights. The situation is illustrated in Fig. 4.2.

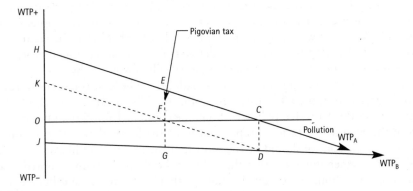

Fig. 4.2 The Pigovian tax.

The control authority, a central or local government body concerned with the control of pollution, observes that A is producing pollution of OC and that this is inefficient; the pollution is excessive. It therefore subjects the polluter (A) to a tax on its emission of pollution. For each unit of pollution produced A must pay a tax equal to EF. The tax shifts A's demand curve for pollution, its WTP, downward by the extent of the tax. The new net of tax WTP curve is KF, the dotted line in Fig. 4.2. Faced with the tax the polluter reduces pollution to OF, since at F its WTP, net of tax, is zero. OF is of course the social optimum and the tax moves the society to the optimum degree of pollution.

But if a Pigovian tax moves the economy to an efficient point, to the Pareto optimum, it does not yield a Pareto improvement in doing so. Two parties are better off as a result of the tax: B by $GDCF$, her willingness to pay to reduce A's pollution from C to F; and the control authority, which receives the tax revenue $KFEH$ (equals the tax rate, $EF \times OF$, A's pollution). However the decision-maker, A is decidedly worse off. Not only has he received no compensation for the losses he incurs in reducing his pollution FCE, but he also has to pay the tax with the consequence that the consumer surplus he receives on his consumption of the optimum level of pollution, OF, is reduced.

The tax is a transfer of consumer surplus from the polluter to the control authority. It does not, therefore, alter the social surplus achieved by the reduction in pollution to the optimum. This remains as before $GDCF - FCE$. The tax therefore has allowed society to realize the social surplus, it has achieved the potential Pareto improvement, but it has not brought about an actual efficiency gain. The gainers *could* compensate the losers and still be better off but as it stands the tax makes the polluter worse off and other parties, B and the control authority, better off. *Thus a Pigovian tax corrects the externality but does not result in an efficiency gain for society.*

What the tax does is to redistribute welfare. B gains $GDCF$, the whole of her WTP to reduce the pollution. Thus she has in fact captured more than the social surplus. If the authority is to turn the situation into an actual Pareto improvement, it must not only return the tax revenue to A but in addition it must tax B by FCE and give that to A as well, thus leaving A neither better nor worse off than he was before the tax was imposed. It must furthermore return this money in a way that does not cause the polluter to change his decision. This is what is called a *lump sum payment*, a payment which the recipient does not anticipate and which therefore does not alter her behaviour in the market concerned. Whether it is possible to devise true lump sum compensatory systems is a moot point but it is an issue that we will not pursue here.

Even were this lump sum payment made, the whole of the social surplus would accrue to those who suffer the pollution. The reader might think that this is altogether just: that polluters should be controlled and should not benefit from those controls. Exact compensation for the losses they suffer in consequence of the control should be the most that the polluter could expect. Some might go further and argue that the polluter should not be compensated at all and that they should suffer for the damage that they impose on others. I consider this issue again in Chapter 5. For the present four points need to be made:

1. Pollution is an externality; it is an accidental or incidental consequence of some unit pursuing its own economic interest. The model does not allow that polluters are deliberately trying to damage the welfare of the sufferers. Were there deliberate intent a different model would be needed since the underlying assumptions would need to be changed; but that position is hardly tenable except in pathological cases. Instead it might be argued that, although not setting out with the intention of damaging others, polluters are showing reckless disregard for their interests. This view undoubtedly does underlie public perception of, and some aspects of public policy on, pollution and is considered later in this and the following chapter. However it does not fit readily into the neoclassical view of pollution as a Pigovian externality.

2. Within the Pigovian model the issue is dealt with (but inadequately) by the issue of property rights. In our discussion of the Pigovian tax the property rights were vested with the polluter and it this that conveys the presumptive right to compensation. Thus the view that polluters deserve no compensation derives from the belief that property rights should lie with the sufferers. This is examined below.

3. The neoclassical view of the world sees the economy as existing to realize the mutual benefits of voluntary cooperation through a system of markets. All actors in the economy, the individuals or the households they comprise, have equal rights in this system and the efficiency criteria we have derived depend on this. Efficiency is raised only when some benefit from changes in activities and none lose. The welfare of polluters in this view is equally as important as the welfare of those who suffer the pollution.

4. In addition to allocative efficiency encapsulated in the Pareto criterion for a welfare improvement, a second component of welfare is recognized, namely distribution. Distribution relates to income levels and the presumption is that more equal distributions are to be preferred to less equal ones. But there is no presumption in the model that the polluters are rich and the sufferers poor. This is perhaps an understandable assumption if the picture is of an industrialist polluting the atmosphere of a neighbouring housing estate, but if the pollution comes from a neighbour's loud music things are not so clear. The model is supposed to cover both of these cases.

Property rights, Pigovian taxes and the polluter–pays principle _____

As an instrument for eliminating excess pollution in the presence of transactions costs, at least if it is to do so in a way that meets the Pareto conditions for welfare improvement, the Pigovian tax runs counter to a belief that property rights should lie with the sufferers and not the polluters. What then should be done if property rights lie with the sufferers but transactions costs prevent efficiency gains? An example might be the previous case of night flying at an airport disturbing the rest of nearby residents. We can again use our basic diagram which, modified, appears as Fig. 4.3.

If sufferers have the property rights, their preferred position, the starting point for analysis, would, as we saw previously, be *O* not *C*. If this were to prevail then there would be too little pollution, not too much, with society losing the benefits of the products that the polluter would produce were he permitted to pollute; there would be no night flying and society would suffer because some people benefit from the opportunity to fly at night. As before we denote the polluter as A and the sufferers as B. The public goods aspect of this problem is the cause of the transactions costs but is otherwise, for the present, ignored.

To persuade B to opt for the social optimum the control authority would, in this case, have to offer them a Pigovian subsidy. To economists a subsidy is simply a negative tax. The Pigovian subsidy is a subsidy paid to B for every unit of pollution that A produces. It is equal to *FG* (= *EF*), the WTP at the optimum. This subsidy raises B's WTP curve to the dotted line *LF* in Fig. 4.3 and turns the 'bad' into a 'good' for pollution levels to the right of *C*. The sufferers, B, faced with the subsidy will choose to permit *OF* pollution, since at *F* their negative WTP is just equal to the subsidy received. Thus the Pigovian tax (in this case a negative tax or a subsidy) again overcomes the transactions cost problem and moves society to the optimum. As before, therefore, we have a potential Pareto improvement. But is this actualized? And how is the social surplus distributed?

In fact in this case both A and B are better off as a result of the subsidy. The social surplus is, as before, *OFEH – JGFO*. The polluter, A, captures his entire WTP of *OFEH*, which is more than this surplus. The sufferers, B, have their loss of *JGFO* more than offset by the subsidy they receive, *JGFL* and are better of by *OFL*. But of course the total gains cannot be greater than the social surplus and indeed must equal it. Reconciliation is through the loss experienced by the control authority, which has to pay the subsidy *JGFL*. Thus the control authority has to find the revenue for the subsidy. However in this case it has a straightforward way to do so; it can impose the Pigovian tax on the polluter. This will not affect the pollution since A is not the decision-maker; he does not have the property right. But in any event A will willingly pay the tax since he will still be better off than if he was not permitted to pollute at all. His gain,

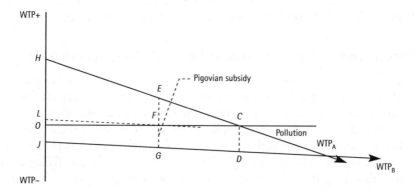

Fig. 4.3 The Pigovian subsidy.

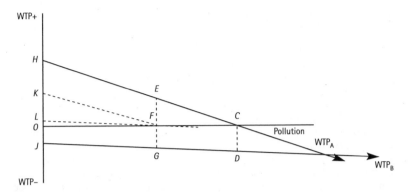

Fig. 4.4 The Pigovian tax and subsidy.

net of tax, will be *OKF* (Fig. 4.4) made up of *OFEH* minus the tax paid. The control authority will break even, if we ignore, as we have done throughout, any costs of administration. In effect the control authority is taxing the polluter and transmitting the revenue to the sufferer. The social surplus is now divided between A and B with A receiving *OFK* and B getting *OFL*. Readers can work out for themselves that *OFK + OFL = OFEH – JGFO*, i.e. that the sum of the gains equals the total social surplus.

The fact that a Pareto improvement is easier to achieve with the property rights residing with the sufferer and that subsidies do not have to be paid to the polluter are the reasons why countries who are members of the Organization for Economic Cooperation and Development (OECD) have subscribed to the polluter-pays principle (PPP). Under the PPP the signatory countries, in effect all the Western developed nations, agree to conduct their pollution-control policies so that the property rights lie with the sufferers. Under PPP the sufferers are not subsidized but the polluters are taxed and the state uses the revenue in ways that benefit its citizens (or in theory anyhow).

Of course firms that have been polluters for a long time and have an established *de facto* right to pollute would undoubtedly be made worse off by the implementation of PPP. In recognition of this many governments apply only an attenuated version of PPP, whereby existing polluters are licensed to pollute at agreed levels and are only liable to financial sanctions should they exceed their consented levels. In other words they are not subjected to Pigovian taxes.

Summary

- The economic theory of pollution sees pollution as an uncorrected externality.
- In the absence of policies on the matter, property rights lie with the polluter.
- The optimum level of pollution is the level of pollution at which gains from trade are eliminated. This is the point at which the willingness to pay of the sufferers for reductions of pollution is equal to the compensation that would be required of the polluters.
- The optimum is independent of the allocation of property rights being determined by the willingness to pay functions of the parties to the problem.

- This optimum will not be zero because the polluter benefits from creating pollution.
- The economic optimum level of pollution bears no necessary relation to any biological notion of the absorptive capacity of the receiving medium.
- If the optimum is achieved through negotiations between the parties the distribution of the resulting social surplus is indeterminate.
- In general a negotiated solution will be prevented by transactions costs. These can include disputes over property rights, conflicts over the distribution of the social surplus and lack of trust between the parties. Where there are many sufferers or many polluters there may also be high costs of negotiation and free riding.
- In the presence of transactions costs a control authority may bring about a solution by the use of a Pigovian tax.
- A Pigovian tax will achieve the optimum level of pollution but it will not yield a Pareto improvement (efficiency gain) since it will make the polluter worse off. To achieve a Pareto improvement the sufferer must be taxed and the polluter compensated for his losses by a lump sum tax.
- A control authority may also determine the allotment of property rights. If it gives these rights to the sufferer then a combination of a tax on the polluter and a subsidy to the sufferer will both achieve the optimum and provide an efficiency gain. This is an argument for the polluter-pays principle.
- However, except for new sources of pollution, removal of de facto property rights from polluters makes them worse off. In consequence only an attenuated version of the principle is applied to established polluters.

Chapter 5

A critique of the Pigovian model of pollution

The last chapter presented the economic model of pollution as an example of a Pigovian externality, compounded by transactions costs. These transactions costs could include the public goods problem of free riding in situations where there are many parties involved, but might also embrace imperfectly specified property rights. This model allowed the identification of an optimum degree of pollution. This was not zero and was defined as the pollution that would result were the transactions costs removed and the parties allowed to freely bargain. Given obstacles to bargaining in the form of transactions costs, the way was open to a control authority to bring about the optimum degree of pollution by the imposition of a Pigovian tax. The parties' valuation of the pollution allowed the appropriate tax to be specified.

In its essentials neoclassical environmental economics applies this model not only to issues of local pollution but also, with modifications, to issues of transboundary and global pollution such as acid rain, damage to the stratospheric ozone layer from chlorofluorocarbons (CFCs) and to the problem of global warming. By analogy it is also used for problems that would not be seen as pollution in the ordinary sense such as loss and damage to wildlife and wildlife habitats. The notion of an optimum state of the environment is an outgrowth of the Pigovian model. Since it is a central plank of economists' thinking on the environment it is important to subject the model to a critical analysis. That is the function of this chapter.

While in its details this model seems arcane and complicated its essence can be captured in a simple parable. This will help us to see its limitations.

A local neighbourhood has its rest disturbed by the habit of one of the households of playing loud music for lengthy periods at inconvenient hours. The immediate neighbour goes to the head of the offending household and offers to pay her if she will reduce the volume of the music for part of the time. They are both reasonable people and a bargain is struck. Music is curtailed late at night and on Sundays in return for a weekly payment. The result is that the neighbourhood is quieter. One person is poorer but has more peace and quiet. The noisy household has its freedom restricted but can use the money to take its members to venues where loud music can be enjoyed. The limits to the bargain are determined by the quiet household's willingness to pay for silence and the compensation that the noisy household requires for restricting its domestic music. Bargaining will cease at the point where the latter exceeds the former and the resulting degree of neighbourhood disturbance is the optimum degree of (loud music) pollution. This result has a moral justification in that it is the outcome of an agreement freely entered into. It has, therefore, the virtues of the marketplace.

It is easy to find reasons why this agreement might not be reached. Most obviously the quiet household might have asserted its right to silence and instead demanded compensation for the noise it suffered. If the noisy household accepted the dominance of the right to silence over the right to music then the bargaining could still have taken place but the money would have flowed in the opposite direction. Most likely, however, its response would be to assert its right over that of its neighbour. Bargaining could therefore only take place if the issue of property rights were resolved. Given conflict this could only be determined by a third party with the power to adjudicate. In line with our parable this would probably be the local authority with its powers to make by-laws. But if the property rights are disputed in this way it is unlikely that sufficient degree of trust would exist for the bargain to be struck. The local authority might therefore be called upon to enforce a solution to the dispute.

In doing so it would need to recognize that there were other affected parties. Surrounding households also suffer the pollution; they will benefit from any bargain struck because the disturbance produced by the noisy household is non-rival – anyone in the immediate vicinity is affected. The quiet household is left to do the bargaining because the other affected households free ride on the outcome. Were they prepared to reveal their preferences, the resources available for negotiating a reduction in disturbance would have been greater and the optimum degree of noise pollution would have been lower. Thus the local authority could intervene to produce more quiet and less noisy music than would happen otherwise. It is therefore correcting a market failure. It can produce the optimal outcome by levying a Pigovian tax (a noisy music tax) on the noisy household. To determine its tax and achieve the optimum outcome it has to persuade all the affected parties to reveal their valuations of the problem. How this is to be done, if at all, is set aside for the present.

This simple parable contains the essence of the Pigovian model of pollution. Let us spell out its implicit assumptions:

1. The transactions are unambiguously defined in time and space and the polluters and sufferers are temporally coincident (i.e. they exist at the same time).
2. The source of the pollution is known to the parties concerned (i.e. the polluters are clearly and unambiguously defined).
3. The sufferers perceive the pollution as a nuisance.
4. Having done so they are able to evaluate its effects on themselves and express them in monetary form as a willingness to pay (WTP) or a willingness to accept (WTA) compensation.

The fourth assumption is considered in greater detail in Chapter 11 on putting monetary values on the environment. It is clear that the individuals trading in markets must possess this capacity to some degree if markets are to operate efficiently. Neoclassical economics assumes a high level of skill as portrayed in continuous (and continuously differentiable) demand curves. While markets plainly could not work if consumers were unable to decide whether goods were worth £100 or 100p, some capacity for valuation less than the neoclassical maximum prevails in practice. Economics generally takes this capacity for granted and does not ask where it comes from. One might ask whether this ability is in some sense innate or is learned behaviour. If the latter, then the degree to which it can be applied outside of familiar contexts and into areas such as the valuation of pollution and the surrogate markets considered in Chapter 11 is open to question. For the present we assume that sufferers from pollution possess the capacity to express their suffering in monetary form.

Of the other three assumptions, some or all of them fail to hold for a wide range of real-life situations of pollution and do not hold for all of the serious pollution problems that constitute the reality of what might be called the current environmental crisis.

Temporal coincidence

Much of the current concern with the environment is about the impact of current activities on future generations. Indeed the inter-generational problem gets to the heart of the environmental crisis since the major global problems of global warming and depletion of the genetic stock are of this kind. Current generations may actually benefit from these effects. The much reported quote from a farmer that 'If this was global warming I want more of it' is believable if apocryphal; and in any event it is far from clear that the costs of preventing the problem are justified to current generations by the perceived benefits. Clearly sufferers yet unborn cannot negotiate with current polluters over the permitted amount of pollution and any bargaining must in a sense be hypothetical. But if there is to be even hypothetical bargaining over inter-generational pollution, future generations have to bargain through agents. These agents, furthermore, cannot be appointed by their principals since the principals, future generations, do not exist. The issue of inter-generational agency, who speaks for future generations, is one that is examined when the notion of sustainable development is discussed in Chapter 14. For the present it is sufficient to note that inter-generational effects pose difficult problems for the view of pollution as a Pigovian externality.

Another aspect of this problem arises when the sufferers are current populations but the polluters existed in the past. The build-up of greenhouse gases in the atmosphere is cumulative and has been taking place since the Industrial Revolution. Obviously the sufferers cannot bargain with the dead, so the burden of control falls on the polluters currently in existence. The impact of the activities of these polluters is, however, time dependent; it is a function not simply of their current levels of discharges of polluting substances but of what has gone before. Again the timeless metaphor of the Pigovian externality is strained by this complication. How the sufferer perceives and values the impact of the polluter's activity depends on its historical context. This is not much like the case of loud music.

Identity of polluters

In the Pigovian parable there is no problem in deciding who is the polluter. Reality can be somewhat different. There are several points to be made here:

1. Many pollutants only become such above certain concentrations in the receiving medium. Below that level they may be harmless or even environmentally benign. An example would be nitrates in water sources. If nitrate discharges from a series of sources are within the absorptive capacity of the water source and are considered safe and an additional source comes into being that pushes the total nitrate discharges to a level at which pollution is deemed to be occurring, who is the polluter? The new source? Or do all of the pre-existing sources become polluters as well?

2. The concentration at which substances are considered to be a problem is often a matter of scientific dispute and thus is liable to change over time. Are those responsible for the discharges always polluters or do they become so only when it is decided that the substance causes damage?

3. Pollution is often classified into point pollution, where there are a number of identifiable points at which pollution enters the receiving medium, and non-point pollution where the source is diffuse. Run-off of chemicals from arable agriculture is a case of the latter. In this case the polluters are an unspecified subset of the farming population.

4. Pollution can arise as a result of the cumulation of substances that were not recognized as pollutants when they were emitted to the environment. Here it may be possible to identify the polluters (they may of course be dead) but they only become polluters after the event.

5. Several substances deriving from different sources can contribute to one problem. This can be illustrated by the problem of global warming. There are a large number of so-called greenhouse gases that contribute to the phenomenon of global warming. Policy debate concentrates on emissions of carbon dioxide, CO_2, derived from the combustion of fossil fuels in electricity generation and from vehicle exhausts. There are of course many other sources of CO_2 emissions including plant and animal respiration. Large quantities of carbon are 'tied up' in living trees. Owners of animals and foresters cutting down trees are not seen as polluters. CO_2 is one of the least powerful of greenhouse gases in terms of its impact on global warming but is important because of the volume of emissions. A much more powerful greenhouse gas is methane, which is released from the decay of organic matter and is emitted in considerable quantities from animal husbandry. The decision to designate certain sources of emissions of CO_2 as pollution, to ignore others, and to ignore also methane emissions may well be an efficient strategy for controlling global warming but it does not fit with the Pigovian notion of the clearly specified polluter.

In summary in many real-life situations *the designation of what emissions shall constitute pollution and in consequence who shall count as a polluter is a social decision.* The decision is not self-evident and cannot be ignored in the analysis of the problem of pollution.

Pollution as a nuisance

If the preceding comments pose problems for the interpretation of the Pigovian model in practical contexts, the next point is damning. In many and indeed in almost all the important cases of pollution the sufferers do not perceive the pollution directly at all. It is drawn to their attention by a third party who identifies it statistically, e.g. as an increased incidence of mortality or disease, or conceptually as a problem on the basis of scientific theory.

For this class of cases pollutants are discovered or invented; they do not simply exist and are not perceived directly by sufferers. Thus, if I am worried about lead in petrol it is because someone has told me that when released into the environment as a result of combustion it is absorbed into the human body and can damage the brains of children. I do not observe this and indeed am not competent to assess the evidence on which the problem is based.

A similar situation prevails for most discharges into water (e.g. nitrates, pesticide residues, heavy metals) and for many discharges into the atmosphere. It holds for the classic forms of global and trans-boundary atmospheric pollution. Thus nobody observes the damage from greenhouse gases and the effects are a matter of scientific hypothesis, with the evidence open to alternative interpretations and still, to a degree, a matter of dispute. If the evidence on damage to the ozone layer is subject to more consensus it is still the case that, in so far as current populations are at risk, they do not directly observe the process.

Our Pigovian parable related to an aural disturbance, loud noise. One might construct similar stories for offensive smells and visual pollution (black smoke, soot on washing hung in the garden to dry). But there is a wide range of hidden pollutants that are potentially much more dangerous to human health, flora and fauna, and indeed the capacity of the atmosphere to support life, where the senses give no warning. Even with sensible (detected by the senses) pollution it is often the unobservable effects that are of greatest concern. Thus particulates in the atmosphere from fossil fuel combustion cause visual pollution but are a factor in repiratory diseases.

If sufferers are to formulate a WTP for these hidden pollutants then they can only do so on the basis of information provided to them. Their WTP will then be determined in part by the nature of this information; but only in part since they will not be passive recipients of the process. Their response will depend on their perception of the status of the provider and on her motives for provision. Thus the information that nitrates in water can cause stomach cancers, or lead in petrol damage the brains of children, is received differently if it is reported as the outcome of independent research by uncommitted scientists, than where the research is sponsored by the purveyors of de-nitrification equipment and unleaded petrol.

WTP will also depend on the objectives of the recipients of the information since they may use that information strategically. Strategic behaviour is severely circumscribed in our Pigovian parable, confined to negotiation over the division of the social surplus and free riding over the public goods aspect of the externality.

There usually is no one correct set of information on a particular problem since research is open to alternative interpretations and is rarely free from methodological doubts. Indeed the very existence of the effects of some discharges into the environment and hence whether these discharges qualify as pollution remains a matter of scientific dispute. This was the case for a number of years with global warming and the issue of feedback effects is still unresolved. The consequences of specific rates of global warming remain an area of controversy. Even where there is consensus, risk factors are uncertain. It matters for the evaluation of sufferers whether the risk of stomach cancers from excessive ingestion of nitrates is 1 in 100 000 or 1 in 1000! In fact there is considerable doubt that there is any such effect at all.

The issue of risk introduces another facet of the problem. The sufferers' evaluation of the information depends on their understanding of it. Mathematical probabilities are one area of difficulty here. Understanding is not a problem in the Pigovian parable.

Thus for many forms of pollution, WTP depends on the intervention of third parties; not, as in our parable, to make the market work, to overcome transactions costs and cause the sufferers to reveal their preferences and to make those preferences effective, *but to create the market*. Third parties are not simply facilitators, they are active participants. Any WTP depends then on the dynamics of interaction between the providers of the information and its recipients. There is no single unique set of information that can be provided; no 'right' description of the problem even within the limits of current understanding. It follows that there is no unique willingness to pay on the part of the sufferers. The consequences need emphasis:

- Since the optimum degree of pollution depends on the sufferers' WTP, in a wide range of cases, and in almost all the cases that environmentalists are concerned about, *there is no optimum degree of pollution* in the Pigovian sense.
- Since the Pigovian tax is a tax that achieves the optimum degree of pollution, in these cases *there is no Pigovian tax* either.
- In this wide range of cases, *pollution does not exist as an obstacle to the efficient operation of the competitive market; it is discovered or invented.*
- If, in these circumstances, a WTP is elicited from those identified as sufferers, its meaning is unclear. It has no unique status and certainly not the status accorded it in the Pigovian model. At best WTP can only apply as a short-hand for something radically different from that illustrated in the parable. This matter is examined further in Chapter 11.

One might be inclined to argue that, for this large class of cases, pollution is not properly described as a form of market failure at all. If new knowledge leads to something being deemed as pollution, the market has not failed since it can scarcely be expected to allocate resources efficiently to something which has not been discovered. The proper terminology is that of technical innovation and the market failure entailed is the public goods problem. What we are seeing is technical progress in environmental quality. The new good, improved environmental quality and safety for current and future generations, has the characteristics of a public good. As with all public goods the market will undersupply it and the issues then are how much to supply, how to supply it, and how to cover the costs of supply. The appropriate tool of economic analysis for these purposes is cost–benefit analysis, which is considered in Chapter 10.

This view of pollution therefore offers a somewhat different perspective on the polluter-pays principle (PPP). Consider the situation of an individual or, more realistically, an industrial establishment that has been discharging a substance into the environment for an extended period of time in the belief that it is harmless. At a certain point society discovers or decides that this activity is polluting. An example might be a 'clean' combustion process that emits to the atmosphere only small amounts of known pollutants such as sulphur dioxide and particulates but which in consequence emits substantial quantities of CO_2. If the PPP is applied to this newly identified pollutant and the company is forced to curtail its emissions without compensation, or is required to pay a tax on them, it might be thought to have a grievance.

One might say that the application of PPP to new sources of known pollutants is reasonable, since the operators of those sources experience no loss; their costs simply are higher than would have been the case had PPP not operated. Equally, where we have emissions that have long been recognized as pollution but have not previously been controlled, there is a case for PPP also; the polluters are simply being made to face up to the social costs of their activities. But the discovery of pollutants is different from this; social costs are discovered along with the pollutants. In these circumstances PPP does not obviously accord with conventional notions of natural justice and its application places all of the burden of social adjustment in the face of new knowledge on those who, in the social interest, have to alter their behaviour.

There are possibly three justifications for the insistence on PPP in the circumstances envisaged:

1. Distribution. The polluters are richer than the sufferers and it is therefore equitable that they should bear the costs of adjustment to the new situation. Of course, while it may be true that the polluters are in general richer than the sufferers, not all of the rich will be polluters, and it is not obvious why, say, industrialists should pay and bankers should not. The response to that objection might be that it is reasonable that the rich should carry the risk of meeting such adjustments. Today it is the turn of the owners of large combustion plants to pay the price of environmental progress; tomorrow the costs may fall on stockbrokers or bankers.
2. This view is closely related to another, namely that industry is the sector of society that takes the risks and reaps the profits of industrial progress and any pollution discovered is simply an example of the risks and the progress. In any event, industry is of course only the proximate destination of the

measures to control discovered pollution. The costs will be passed on in part, if not in whole, to the consumers of the products.

3. PPP may be justified on a cost–benefit analysis as being a principle that leads to the socially efficient means of dealing with the problem, in that it meets the objective at least cost to society as a whole. Cost–benefit analysis is concerned with potential Pareto improvements and not actual ones, so the issue of who actually pays is irrelevant provided that the gainers can compensate the losers and still be better off. Further discussion is deferred until cost–benefit analysis is discussed.

If there is no economic optimum degree of pollution in the Pigovian sense, society still operates with pollution targets and pollution standards. The absence of a Pigovian optimum is no serious inconvenience and there is nothing mysterious about how pollution targets are set. They are set on the basis of scientific evidence or hypothesis mediated through the political process. Targets are fixed to avoid deleterious effects on health and take account of the needs of future generations. Depending on the nature of the problem these may be seen as ceilings or desired levels, mandatory or advisory. Because of their dependence on scientific knowledge they are always provisional and may be expected to evolve.

Thus in practice the potential conflict between the level of discharges into the environment that the sufferer is willing to pay for and the pollution that the environment can absorb, i.e. the lack of any logical connection between the optimum degree of pollution in the economic sense and biological or chemical notions of optima or maxima, is not a problem. Scientific requirements play a role in determining targets for pollution control whatever the market may dictate. Pollution control is a conscious social and political process that should not and cannot be left to market forces.

This issue of pollution targets and the optimum degree of pollution is discussed further as an aspect of environmental cost–benefit analysis, but first we turn to consider the issue of the choice of policy instruments to control pollution.

Summary

- The neoclassical theory of pollution, based on work by Pigou, sees pollution as a nuisance and hence as a cause of conflict between the polluter and the sufferers. The theory defines an optimum resolution of that conflict, the optimum degree of pollution, and says how a third party might bring about this resolution if the parties cannot resolve the matter themselves. This insight underlies the commitment of members of OECD to the PPP.
- This insight has little relevance to the complex problems of environmental pollution that arise with sustainable development for a number of reasons. Many of the problems are not directly perceived by a set of sufferers as nuisances; indeed they are not directly perceived at all. The sufferers can be future generations and there may be problems of defining who are the polluters. Uncertainty and scientific dispute surround the passage from cause to effect.
- In these circumstances the notion of an optimum degree of pollution will have an entirely different meaning from the Pigovian notion, if indeed it has one at all. The view of a control authority bringing about a resolution of the conflict by the use of a Pigovian tax imposed on the polluter is also lost. This leaves the issue of how pollution is to be controlled wide open; it is the subject of the next chapter. It also brings into question the appropriateness of the PPP. We see subsequently that the PPP can cause problems for protecting the interests of future generations and hence for achieving the aims of sustainable development.

Chapter 6

Choice of instruments for pollution policy

This chapter is concerned with what has traditionally been a central issue for environmental economics: the choice of policy instruments to achieve environmental objectives. I concentrate on instruments for the control of pollution, although much of what is said will be applicable to other environmental objectives such as the protection of wildlife or landscapes and, in consequence, is referred to in subsequent chapters. Following the argument in Chapter 5 the discussion is concerned with instruments to achieve pollution control targets and not with attaining some optimum degree of pollution. As a consequence a pollution target is a constraint on economic activity and the level of pollution is not a policy variable. Abandoning the concept of an optimum pollution level simplifies the problem by removing the WTP of the sufferer from the analysis and there is instead an exogenous pollution target that the control authority has to achieve. Beyond that simplification, abandoning the notion of an optimum degree of pollution makes little difference to the substance of the following discussion.

As in Chapter 4 I assume that the producing unit has a demand for pollution. An increase in its pollution level saves it costs on pollution control. These costs may take the form of workers employed on pollution-control activities and capital equipment designed to control pollution. An increase in pollution then allows it to reduce employment and use less sophisticated and expensive capital equipment. Alternatively (or additionally; these are not exclusive alternatives), controlling pollution may place constraints on production so that an increase in pollution allows higher output and higher sales.

As in Chapter 4 also, I assume that the benefit to the polluter from increasing pollution declines as the level of pollution rises, to the point where the polluter is spending nothing on control activities and its production is unconstrained, when the benefit it derives from further pollution is zero. This benefit function is the polluter's willingness to pay (WTP) for pollution; however, to emphasize the point that I am not considering the possibility of negotiation between polluters and sufferers I change its name to the marginal benefit of pollution (MBP) curve. It is of course simply WTP_A in Fig. 4.1 and represents the polluter's demand for pollution.

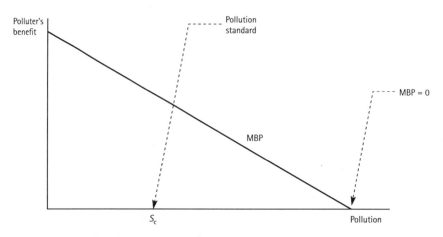

Fig. 6.1 The conventional problem of pollution control.

In the absence of any constraints the producer will choose to pollute to the point where MBP = 0. This is shown in Fig. 6.1. For simplicity MBP is shown as a straight line but the argument of course only implies that it is a declining function of the level of pollution.

The pollution target of the control authority is shown simply as a level of pollution, a point on the horizontal axis, designated here as S_c. A pollution-control problem exists if S_c lies to the left of the point where MBP = 0 since in that case, in the absence of some means of controlling the polluter's activity, pollution would be greater than is deemed acceptable.

Definition of an instrument

An instrument for pollution policy is any device that causes the polluter to comply with the wishes of a control authority, i.e. that causes him to reduce his pollution to the target level S_c. In Chapter 4 we considered one such instrument, the Pigovian tax, but that instrument depended on the concept of an optimum degree of pollution, which has been rejected. There are a large number of instruments that could be used for pollution policy and the choice of instrument in practice, together with the specific details of its design, will depend on a large number of factors that are specific to the particular problem under consideration (Box 6.1).

Box 6.1 Factors determining instrument choice

1. The nature of the discharge: gaseous, liquid or solid form.
2. The receiving medium: air, water or land.
3. Substitutability between receiving media and transport mechanisms, e.g. can the discharge be in either liquid or gaseous form and released into either the atmosphere or the water system?
4. The toxicity of the polluting substance and its persistence in the environment.
5. Detection technology: is pollution detectable by inspection, measurable by installed automatic instrumentation, or does it require laboratory analysis of samples of discharge flows or of the receiving medium?
6. Whether the pollution is sensitive to location of entry into the environment (thus the location of discharge of greenhouse gases is irrelevant but the effects of specific volumes of discharges into rivers varies with location and timing).
7. Whether there are identifiable discharge locations, i.e. whether it is point or non-point pollution.
8. Socio-legal factors: some instruments may be socially unacceptable and/or legally unenforceable.

Some combination of these factors probably determines the choice of instruments in many real-life cases so that in practice the range of instrument choice is severely constrained. The text discussion is concerned with broad issues that apply to instrument choice in situations where choice is possible.

In this chapter I consider only the broad arguments about types or classes of instruments. The types of instruments examined are as follows:

1. command and control;
2. pollution taxes;
3. marketable permits; and
4. pollution subsidies.

We are concerned with what is known as point pollution, where the identity and location of the polluting sources are known and with discharges either in gaseous form to the atmosphere or in liquid form to bodies of water. Non-point pollution and the issue of solid waste disposal are the subjects of Chapters 7 and 8. The issue of trans-boundary pollution, where pollution crosses national boundaries or is of global impact and cumulates through time, is left to a later chapter. The issues considered here may be thought to concern how to control the factory that is polluting a river or discharging noxious gases into the local atmosphere

Traditionally the major debate on instrument choice in the economics of pollution has been over the relative merits of economic instruments and command and control, with the former represented by pollution taxes and the latter by pollution standards backed up by the law. Of late, however, that debate has become blurred for several reasons: the invention of a wider range of instruments; more sophisticated analysis of how traditional instruments function; and new considerations introduced by sustainable development and the recognition that environmental damage caused by pollution may be irreversible. I begin with some basic points of definition.

Command and control relies primarily on the law to achieve its objectives. The polluter receives a license or 'consent' to discharge pollutants into the receiving medium at specific rates and it is a legal offence, punishable typically by a fine, to exceed the conditions specified in the consent. A typical command and control instrument is the system to control the pollution of controlled waters, essentially all rivers, lakes and coastal waters in England and Wales. The control authority, the Environment Agency, issues consents to industrial premises and sewage works for discharging into controlled waters and its is an offence, punishable through the courts by fines, to discharge polluting substances in breach of consent conditions. The consent conditions relate to the volume, chemical composition, temperature and timing of the discharges.

An economic instrument is one designed to work through markets, i.e. it utilizes the economic system to achieve the objectives. The classic economic instruments are taxes and subsidies. Polluters are either taxed on the pollution that they produce or are offered subsidies for not polluting or for introducing pollution-control equipment.

These economic instruments might be termed *command economic instruments* in that taxes are collected or subsidies paid by some central authority, which may be a national, a regional or a local authority; what matters is that the authority has fiscal powers, i.e. the power to levy taxes. The alternative economic instruments are what are called *marketable instruments. The polluter is given a permit to emit a certain quantity of pollution which he has the right to sell or otherwise transfer to someone else.* Marketable instruments convey limited property rights on to polluters: the right to pollute within the limits of the permit. In doing so pollution is turned into a private good that has value to the polluter since he may transfer that right to someone else. With marketable instruments pollution becomes an excludable good to the polluter: only those with the requisite permit may pollute and the holders of the permit can transfer the exclusive right to that pollution.

Economic analysis of instruments

It is important to distinguish economic instruments as a class of instruments from the economic analysis of instruments. Instrument choice requires an understanding of how instruments achieve their objectives and this is the subject of economic analysis. This analysis is required because neither economic instruments nor command and control instruments are self-policing. Thus with command and control it has to be recognized that people do not necessarily obey the law. There are annually several thousand breaches of consent conditions for discharges to controlled waters in England and Wales. Equally, people and companies do not automatically pay taxes and in fact tax avoidance is a major branch of accounting.

Because they are not self-policing all economic instruments must ultimately be backed by the law. If the system of discharge consents to controlled waters in England and Wales were replaced by pollution taxes, it would have to be made a punishable offence to evade those taxes. But what would amount to evasion would depend on the conditions under which polluters were liable for tax. These conditions would be determined as the outcome of a series of legal battles between the tax avoidance industry and the control authority.

If, alternatively, the discharge consents to controlled waters were converted into marketable permits, it would become an offence to discharge while not in possession of the requisite marketable permits and what that ultimately meant would again be decided through the courts. If there were no enforcement mechanism for marketable permits then they would have no market value since the right to pollute would be non-excludable and the owner of permits would not be able to transfer any exclusive right to possible purchasers.

The need for legal back-up applies equally to the payment of subsidies. If a firm receives a subsidy for installing pollution-control equipment or for reducing its volume of pollution, there must be a threat of sanctions to prevent it from taking the money but not installing the equipment or reducing its pollution.

A threat of sanctions will only be effective if the control authority is able to detect when an offence has been committed. Thus as well as legal back-up all instruments require effective monitoring. Control without monitoring is impossible. This is part of what is meant when we say that no instruments are self-policing.

Both monitoring and enforcement of pollution controls entail costs that are collectively termed *compliance costs*. The incidence of these compliance costs, whether in the first instance they fall on the polluter or the control authority, will depend on the specific details of the instrument. I use the term 'first instance' here because the final incidence of these costs may be different from what appears at first blush. The polluter may be able to pass the tax on to consumers as higher prices or may be forced to absorb it as reduced profits. If the polluter is a multiproduct company the price rises may not be borne by the products giving rise to the pollution and hence may not be paid by the consumers who, in the sense that they are consuming those goods, are benefiting from the pollution. Equally, the control authority may recover its costs through increases in general taxation, some of which will be paid by the polluters. The incidence of the costs of environmental controls is part of a broader problem that is discussed in later chapters but is ignored in this one.

Regardless of the incidence of compliance costs, society has an interest in minimizing their total value since the resources committed to control have an opportunity cost. The level of compliance costs is thus one factor in instrument choice. As such it is discussed below after we have considered the economic analysis of each of the instrument classes.

Command and control

An economic analysis of command and control as an instrument sees it as a peculiar form of tax. As an example consider a factory that has a consent to discharge some pollutant to a controlled water. If it discharges at rates above the consent conditions it will derive some benefit, i.e. at the consent level its MBP is positive. If that were not the case there would of course be no need for pollution control.

But if there are benefits to the factory of ignoring the consent conditions, the presence of the control means that there are also costs arising because it will have committed a legal offence. These costs depend on a number of conditional probabilities:

the probability of detection of the offence $p(D)$;
given detection, the probability of being prosecuted $p(P)$;
given prosecution, the probability of being convicted of the offence $p(C)$;
if convicted of an offence the factory then faces a fine. Call the expected fine $e(F)$.

The cost of breaking the consent conditions is then the product of these probabilities and the expected fine:

$$p(F) = p(D)p(P)p(C)e(F)$$

This situation is shown in Fig. 6.2.

MBP is the polluter's marginal benefit curve from pollution, i.e. the factory's demand curve for pollution. S_c is the consented pollution standard. Pollution is free to the factory up to S_c but if its discharge level exceeds S_c it incurs the risk of the sanction and the expected penalty is $p(F)$. If $p(F)$ is set to $p(F)1$ then the costs of pollution exceeds the benefit and pollution will be kept at S_c. If $p(F)$ is set at the lower level $p(F)2$ then it will be worthwhile for the polluter to break the consent conditions and incur the risk of prosecution, conviction and a fine.

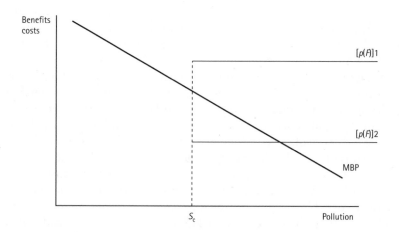

Fig. 6.2 Command and control as an equivalent tax.

Clearly the control authority will want to set the level of $p(F)$ to $p(F)1$ rather than $p(F)2$. It can adjust $p(F)$ upwards in a number of ways:

1. by investing in monitoring and thus increasing $p(D)$, the probability of detection;
2. by increasing its willingness to prosecute offenders, thus increasing $p(P)$;
3. by legislating for a higher scale of fines subject to the discretion of the courts over the precise penalty imposed. A higher scale of penalties would probably, at least initially (until and unless the courts showed they were unable to increase the actual penalties), increase $e(F)$.

The control authority has no control over the conviction rate $p(C)$, which is at the hazard of the law.

From the viewpoint of the factory the control is akin to a tax on pollution of zero for pollution levels up to S_c and $p(F)$ above that level.

This model is probably over-simplified in that it does not allow for feedback. In practice detection of an offence is likely to lead the control authority to increase its monitoring effort and its willingness to prosecute. Conviction for an offence is likely to increase the expected fine that the factory faces for further offences as the courts are harsher with repeat offenders. Thus it might well be that $p(F)2$ is unsustainable and, following the detection of a breach of consent conditions, the feedback effects will push its level up to $p(F)1$. The experience of the control of discharges to controlled waters might lead one to suppose, however, that feedback effects are not particularly powerful since there are a number of well-known persistent offenders.

Pollution taxes and subsidies

In essence a pollution tax is identical to a Pigovian tax except that its rate is set to achieve a predetermined standard and not some presumed social optimum. For the individual firm the tax and the subsidy may shown on the same diagram.

In Fig. 6.3 MBP is the factory's demand for pollution as before. In the absence of any control it will pollute at point A, which is above the desired level, i.e. what would be set as the consent standard S_c under command and control. The control authority therefore levies a tax per unit of pollution equal to CD. This moves the net of tax MBP curve down to the dotted line so that the factory 'chooses' to pollute at S_c.

If, instead of a tax, the control authority chose to pay the factory to keep its pollution within the standard, the size of the subsidy needed is given by the area of the triangle $S_c AB$, which is the total benefit to the factory from the extra pollution $S_c A$.

A subsidy on the installation of pollution-control equipment would be designed to reduce the MBP curve down to the dotted line. The size of this subsidy cannot be shown on the diagram since it would depend on the cost of pollution-control equipment. It would, however, be a one-off payment while the subsidy for reducing pollution, $S_c AB$, would be a continuing one. In order

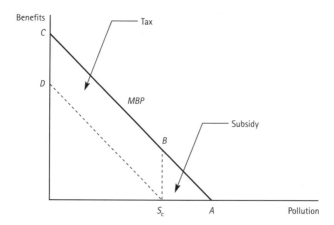

Fig. 6.3 Pollution tax and subsidy.

to compare the two types of subsidy, the continuing subsidy would need to be capitalized and expressed at its present value. The notion of present value is discussed in Chapter 9 on cost–benefit analysis.

One way of financing a subsidy on the installation of pollution-control equipment would be to levy a pollution tax and to recycle the revenue to the polluters in the form of a system of grants for the installation of pollution-control equipment. If successful such a tax–subsidy scheme would self-destruct; the subsidies would fund investment by the polluters that would reduce their future tax liabilities.

Even without recycled subsidies, one possible response of the polluter to the levying of a pollution tax would be to invest in pollution control to reduce tax liability. The tax–subsidy proposal works where the rate of return that the polluter would receive from investment in pollution-control equipment is insufficient to justify the investment in it. In this case the subsidy lowers the capital cost to the polluters and increases the rate of return on the commitment of their own capital. But there is of course no reason to suppose that the subsidy that can be paid out of the tax revenue will raise the rate of return sufficiently to eliminate the need for the tax. If it does then the *social* rate of return on pollution-control equipment is such that the introduction of control equipment is the best option for society. If not then the best social option includes reduction of consumption of the product whose production generates the pollution.

Subsidies to polluters are in conflict with the polluter-pays principle (PPP) and are generally rejected as an option for that, if for no other, reason. The issue of the justice of applying PPP to cases of 'discovered' pollution is discussed at the end of Chapter 5. For the reasons given there we retain subsidies in the range of options to be evaluated. The tax–subsidy alternative returns the tax to the polluter provided that he uses the tax revenue on pollution control. If that is not his preferred option then he is made worse off by the system and tax–subsidy is not in conflict with PPP.

Pollution taxes with many polluters

Figures 6.2 and 6.3 related to the situation of a single polluter. Typically of course there are many polluters and control systems are designed with this in mind. The situation of a pollution tax with three polluters (which will stand for any number greater than one) is shown in Fig. 6.4. The three polluters have different MBP curves shown as MBP1–MBP3. The control authority wishes to achieve a pollution target of P^*. It sets the tax at the level at which the three polluters 'choose' their pollution levels such that $P1 + P2 + P3 = P^*$. Each polluter sets her pollution level such that her MBP curve, her demand curve for pollution, equals the tax rate. The tax is the price of pollution for the polluters and they maximize their benefits where their demand for pollution is equal to this price. Their consumer surpluses from creating pollution are the areas of the triangles made by the MBP lines and the tax level.

If, instead of a tax, the authority was to operate command and control it would set consent standards to achieve P^*. This is also shown in Fig. 6.4. I assume, for simplicity of diagrammatic presentation (the assumption does not affect the argument), that each polluter is set the same consent level S_c such that $3S_c = P^*$, the desired target. In this case the three polluters end up with different levels of MBP, MBP3 being the lowest and MBP1 the highest. *This means that, as a mechanism for meeting the pollution target, the command and control option is Pareto inefficient, since gains from trade are not eliminated.* Were the three polluters allowed to trade their consent levels, subject to the total pollution level meeting the overall target of P^*, they could find mutually acceptable terms of trade. Thus polluter 1 would be willing to pay any sum up to MBP1 for the right to increase her pollution. Polluter 3 would be willing to sell some of her consent for any price above MBP3. Since the maximum that polluter 1 would be willing to pay to buy polluter 3's consent is greater than the minimum sum for which polluter 3 would be willing to sell her consent, the two parties can gain from trade.

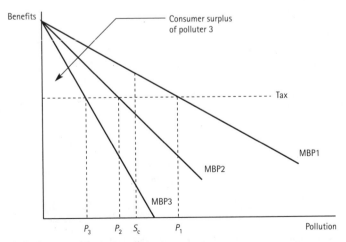

Fig. 6.4 Pollution tax with many polluters.

Gains from trade are eliminated once pollution is distributed between the parties so that their MBP curves are equal, since then the maximum that any polluter is willing to pay for further consents is then equal to the minimum that any polluter is willing to accept as compensation for parting with consents, and any further trades of consents would not be mutually beneficial. That condition is met by the pollution tax and thus *the pollution tax is a Pareto-efficient way of meeting the overall pollution target.*

The existence in a market of a single price that all parties accept and on the basis of which they make their economic decisions is sufficient to eliminate gains from trade. The pollution tax is such a case. A pollution subsidy, which is simply a negative tax, would equally eliminate gains from trade. All economic instruments face all polluters with the same prices and hence all are Pareto efficient. This fact gives us an alternative and more satisfactory way of defining an economic instrument than the loose definition provided earlier, which is open to the objection that with some economic instruments, such as tradable permits, there is no market through which the instrument can operate and *the instrument creates a market.*

Further light can be cast on the nature of the distinction if we consider command and control as a tax (Fig. 6.5). As before, there are three polluters with their MBP curves labelled 1–3. The control authority issues the same consent standard, S_c, to each, such that $3S_c = P^*$. To achieve compliance it must set the controls so that $p(F) > \text{MBP}i$ where $i = 1$–3. This condition is satisfied in the diagram but, as can be seen, $p(F) > \text{MBP1} > \text{MBP2} > \text{MBP3}$, whereas to eliminate gains from trade we would require $\text{MBP1} = \text{MBP2} = \text{MBP3}$. Any instrument that satisfies the equality would be an economic instrument. Command and control does not satisfy the equality but a different requirement, the inequality. Command and control therefore is not an economic instrument.

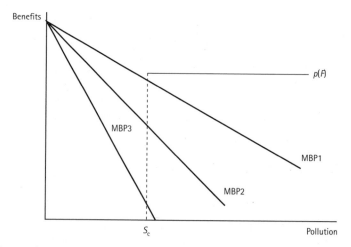

Fig. 6.5 Command and control as an equivalent tax: many polluters.

Choice of instrument _____

This conclusion that an economic instrument is one that is (designed to be) economically efficient is an important one. It has led many economists to argue that economic instruments should always be used for the control of pollution. Governments, by and large, have failed to respond to this more-or-less unanimous conclusion from the economics profession and a number of academic papers have been written by economists attempting explanations of why the advice has been ignored. Therefore it is worth considering the matter a little further. Three points can be made at this juncture:

1. Economic instruments allow a market to determine the distribution of pollution between a set of polluters. This is technically unacceptable in cases where the location of discharges is important. As an example, consider again the system of discharge consents to controlled waters in England and Wales. In setting consent levels for discharges into a specific river catchment the Environment Agency has a water-quality objective specified in terms of a range of biological, chemical and aesthetic criteria. The impact of a given discharge volume of a specified pollutant will vary with the location in the catchment and with other factors such as the time of year. Discharge consents are therefore specified in terms of location, timing, chemical composition, temperature and other factors. Other things being equal given discharges located towards the river estuary and in winter, when flows are high, have a smaller impact on catchment quality than they have in summer and when located in the upper reaches. This degree of control is not achievable with a pollution tax. The polluters with the highest MBPs, who, under the tax, will discharge the most, may be located in the upper reaches and conversely those with the lowest MBPs may be on the estuary. Thus to reach the specified environmental objectives with the tax it would be necessary for the Environment Agency to work to a lower pollution target than would be needed with command and control so as to safeguard the water-quality objective from the uncertainty over how much pollution will occur in each stretch of river. Any efficiency savings from the tax therefore would be offset by the losses from the lower pollution target.

   It is not only discharges to the water system that are locationally sensitive; many atmospheric emissions also have differential locational impacts and inflict greater damage on the immediate surroundings. Thus sulphur depositions from fossil fuel combustion, the principal cause of the trans-boundary pollution problem of acid rain, decline with distance from the emission source. Where local atmospheric quality is a factor in pollution control the authorities will not be indifferent to location. Pollution taxes deprive them of the control on location.

2. Economic instruments allow pollution targets to be met at the lowest aggregate cost to the polluters but this is not necessarily the lowest cost to society since there are other cost factors. Specifically there is the issue of compliance costs discussed earlier. While some writers have chosen to ignore control costs, assuming that they are invariant between instruments or are trivial, this, as discussed below, is not necessarily the case.

3. In order to set the pollution tax the control authority needs to know the MBP functions of each of the polluters. This information is difficult to acquire. Each polluter is privy to its own MBP curve (although they may not perceive or store the information in that form) but they have no incentive to reveal that information to a taxing authority. In fact polluters have an incentive to understate the benefits they derive from pollution, and to overstate the costs of reducing pollution, in order to reduce the tax level set. If the control authority gets the tax level wrong then it will have to adjust it. This could be expensive both for itself and for the polluters if they have invested in control equipment and reorganized their operations in the face of the tax. Thus the information requirements of the pollution tax (and equally of the pollution subsidy) can militate against its use. The obvious strategy to overcome the information deficiency, trial and error, could incur resource costs that offset the efficiency gains of the instrument.

To operate command and control the authority does not need the information on the polluters' MBP functions. It has to set its control variables to ensure that $p(F)$ is above the highest MBP but if it errs on the side of caution and sets the penalties too high there are no losses to the polluters or to itself.

Marketable permits

The information problem is overcome if the control authority issues consent permits, calculated to meet in aggregate the pollution target P^*, and allows the parties to trade them. This is the marketable permits option. Gains from trade will be eliminated once the permits are redistributed so that MBPs are equal for all parties. Thus from the position of equal values of S_c in Fig. 6.5, trading would take place until gains from trade were eliminated, i.e. until the position in Fig. 6.4 was reached. It is obvious that the market price for permits will be the pollution tax rate since that is the price of pollution that eliminates the gains from trade. *Thus, as an instrument, the marketable permit achieves the efficiency gains of the pollution tax but avoids its information problems.* It is not necessary to know the MBP functions of the polluters; it is sufficient that they know them and trade accordingly. Nonetheless there are problems with marketable permits.

Thin markets
Marketable permits are only a feasible instrument in situations where there are many polluters, since extensive trading is required to establish the market clearing price. In situations where there are only a few polluters, or there are other factors inhibiting trade in permits, the market is said to be 'thin'. In thin markets trade may take place at a wide range of prices so that the efficiency gains, which require that all polluters face the same price, may not be realized. If there are only a few polluters and they are in competition with each other in the markets for the products that they sell, then there will be a temptation to use the marketable permit system as a competitive weapon, with companies

preferring to hoard permits that are surplus to their requirements in order to deny them to their competitors, thereby forcing up their competitors' costs.

Initial allocation

As an instrument, marketable permits face a starting problem: how to allocate the permits initially. One approach, known as 'grandfathering' is to give the permits to the producers in proportion to what have historically been their levels of pollution. This is open to the objection that since the permits are marketable assets that may be sold, the recipients are being rewarded for past pollution. The alternative approach is for the control authority to auction the permits but this meets the opposite objection that polluters could face a financial penalty for continuing to do what they are legally entitled to do. The result of the permit auction might be that some polluters failed to obtain their necessary permits and were unable to continue in business. The problem arises because, in contrast with the introduction of command and control, where polluters would be faced with possible financial penalties for polluting beyond consent levels, under tradable permits there is a financial penalty for all pollution including that which would be free under command and control. The auctioning of tradable permits for newly discovered pollution would seem to be particularly unreasonable and grandfathering has typically been the strategy adopted in practice.

Locational pollution

Like pollution taxes marketable permits are unsuitable for control of pollution where the location of the discharge is an important consideration. In consequence most of the currently extant uses concern discharges to the atmosphere rather than to water. One of the most successful schemes has been that for the control of sulphur dioxide and nitrogen oxide emissions from electricity generation plants in the USA introduced as part of the US Acid Rain Program. This scheme allows trading of allowances (the unit is a right to emit 1 ton of SO_2 per annum) between electricity utility companies responsible for about 1000 generating plants. As well as trading their allowances the companies may bank them for use at a future date. This policy has been introduced as part of a phased programme for the reduction of emissions via restrictions on fossil fuel-burning plants (for a discussion of the policy and an assessment of its success see Rico, 1995).

A number of suggestions have been made for overcoming the locational limitations of marketable permits. The simplest has been dividing the market into a series of zones within which location does not matter and between which trading permits or consents is not permitted. The constraint on this as a solution is given by the problem of thin markets. The smaller the zones the less significant is location but the fewer the number of polluters and hence the greater the problem of thin markets. Additionally, zoning increases the costs of monitoring compliance. An alternative technique is to couple the marketable permits with a set of imposed standards and to require that the trading operates within those limits. Thus electricity utilities in the USA are required to meet all local constraints on emissions of SO_2 as well as their obligations under the Acid Rain Program. This solution also runs the risk of creating thin markets, adds to the problem of monitoring and increases compliance costs. A final suggestion is that the consents should have different values depending on the location of the holder. Thus a consent to emit 1 ton of sulphur in one area should only permit the emission of 0.5 ton if sold to a polluter in another area. This solution would require zoning as a matter of practicality and again would increase the risk of thin markets as well as adding greatly to the costs of ensuring compliance.

All of these solutions run counter to another claimed benefit of marketable permits, namely that the monitoring costs are reduced because the control authority can acquire information on compliance by simply observing the movements in the market prices of permits. If the price of permits falls then this is a signal that the supply of them for sale has risen relative to the demand. This could be because the polluters have become more efficient at meeting their pollution control requirements, e.g. because they have introduced new or improved control equipment, or it could be because they are ignoring the need for permits. It is a signal for increased monitoring of the permits. On the other hand, a rise in the market price does not ring alarm bells: it might indicate new sources of pollution entering the market and buying permits but it does not indicate that the controls are failing to work. If the authority can determine the market clearing price, then it can follow a rule that requires it to check for compliance if the price falls below this price but not otherwise. This idea appears not to carry through to real life. As well as private sales between individual utilities, the US Acid Rain Program requires an annual auction of a proportion of allowances to public sector utility companies to ensure that the private sector companies are not denied access to allowances. In 1994, allowances in the 'spot' auction (for allowances that could be used immediately as opposed to the 'advance' auction for allowances that could be used in future years) ranged from $165 to $550!

Other advantages of economic instruments

As well as their advantage of ensuring allocative efficiency, economic instruments are claimed to have two other advantages over command and control:

1. they pose no barrier to new competition;
2. they encourage innovation in pollution-control technology.

Pollution control as an entry barrier

Markets only ensure economic efficiency if they are competitive. Competitiveness requires that producers are free to enter markets if they perceive opportunities for making profits and to leave markets, to cease production, when the opportunities for profit disappear. Pollution control could be a barrier to entry to a market in certain circumstances. Thus if a pollution-control authority meets its targets by subsidizing the existing polluters to install pollution-control equipment a potential new entrant to the industry could be forced to fund the investment in control equipment itself. In this case the potential new entrant would be faced with higher costs than existing producers and thus be placed at a competitive disadvantage. Similarly if targets were met by command and control and existing producers were in aggregate producing as much pollution as the control authority would permit, a potential entrant would be unable to obtain a consent. Thus both command and control and subsidies for equipment installation could constitute entry barriers.

On the other hand, neither pollution taxes nor marketable permits constitute barriers to entry. If a new producer enters a market subject to pollution taxes the response of the control authority will be to raise the tax rate in order to meet the target with the new larger number of polluting sources. The rise in the tax rate will cause all polluters to reduce their pollution levels thereby accommodating the new producer. The rise in the tax rate increases the costs for all producers; it does not disadvantage the new producer. The pollution tax therefore does not constitute a barrier to entry. The effect of a new producer entering a market where pollution is controlled by tradable permits is to push up the price of permits as the new producer bids for its permits. Again all producers are similarly affected and the control does not provide an entry barrier. This will only be provided so that the market is not thin, since in a thin market existing producers may hoard permits and decline to sell them to the new entrant in order to exclude it from the market.

The significance of command and control as an entry barrier is open to question. Pollution control is not based on product market definitions; rather it relates to types of pollution regardless of the purposes for which that pollution is created. A single type of pollution can come from sources producing for a variety of markets and utilizing a variety of technologies. If producers for a particular market use a variety of technologies they may not all be subject to the control. Thus the sulphur controls in the USA are not needed for nuclear and hydroelectric generating technologies even though these are serving the same markets for electricity as the fossil fuel generators that are controlled. Furthermore if the pollution is locationally sensitive and the producers locationally scattered, they may be subject to different levels of pollution control, e.g. with command and control they may have different consent conditions and consented levels of the pollutant; so that, even with economic instruments, they may not be all equally affected by new pollution sources. Pollution control places constraints of location and technology on new competition as it does on existing producers.

Innovation and pollution control

It is frequently argued that economic instruments provide a stimulus to innovation in pollution-control equipment while command and control does not. In consequence, whatever the difference in social cost of different instruments at any point in time, economic instruments are to be preferred since, through time, their relative cost will fall. The rationale for this view is simple: economic instruments (excluding subsidies for the moment) offer a financial incentive for reducing pollution, in the form of a reduction in taxes paid or, with marketable permits, the freeing of an asset that can then be sold. With command and control the effective tax rate, provided the polluter remains within the consented level, is zero; there is no financial incentive to reduce pollution.

This argument is not only simple, it is fallacious. As we have seen, any control instrument imposes costs on the polluter in the form of constrained output of its product and/or expenditure on pollution-control measures. If it did not then there would be no need for the control in the first place since pollution targets would be met without it. It is this cost penalty that provides the incentive for innovation on the part of the polluter. With command and control, innovation in pollution control raises profits by reducing cost; with marketable instruments profits can be raised additionally by sales of permits.

What *is* true is that, while both economic instruments and command and control provide incentives to the polluter to reduce the cost of meeting the constraint, economic instruments provide an *additional* incentive because all pollution, and not just pollution above target levels, is subject to a financial penalty.

The difference that economic instruments make becomes clear if we suppose that among the set of polluters subject to control there are some for whom the controls are unnecessary because they can meet their consent standards without any costs. This subset of the polluters can stay within their consent levels without constraining output and while abating pollution only to the extent that they need to in their own interests. If pollution control is by economic instrument this subset will still face financial penalties; they will still have to pay tax on, or buy permits for, the pollution that they do produce. Under command and control this subset would have no incentive to reduce their pollution; under economic instruments they will.

Whether this additional incentive to pollution reduction is seen as an advantage or a drawback of economic instruments depends on whether the additional investment induced constitutes an efficient use of society's savings. This is important because *resources devoted to innovation in pollution control have an opportunity cost for society; they might instead be devoted to finding a cure for cancers or improving the education system.* If the judgement is that it is an efficient use of society's savings, then whatever instrument is chosen to meet pollution targets should be accompanied by an incentive for innovation in pollution control. This incentive might take a variety of forms. It might be a subsidy for investment in pollution control. It might simply be a programme of planned reductions in future emissions levels so that polluters know that they will have to conform to tighter targets in the future. Such a programme is part of the US Acid Rain Program.

But if there is a need for more investment in pollution control than the existing set of pollution targets will deliver there is no guarantee that the extra investment engendered by the use of economic instruments will be the appropriate amount. It might be too much or too little. It would be socially efficient in other words to take a considered decision on the resources that should be devoted to this investment and find a mechanism, or a set of instruments, to deliver that investment.

The argument for believing that society would benefit from more reduction in pollution than the existing control targets will provide starts from the recognition that the cost of control is a factor in the specification of pollution targets in any event. This is manifest in the UK system of integrated pollution control introduced under the Environmental Protection Act 1990 whereby the Environment Agency, in determining the medium for discharges of controlled substances and the permitted discharge rates, has recourse to the concept of best available technology not entailing excessive costs (BATNEEC). Cost considerations affect both the targets and the timetable for meeting them as, for instance, in discussions on the Second Protocol on Acid Rain where scientific evidence of the reductions needed to keep emissions within the absorptive capacity of the environment were deemed by the participants to be wholly unacceptable on grounds of cost (Sweet, 1994).

The extra investment in innovation produced by economic instruments for pollution control is an accidental by-product of the control system. It arises because, like the Pigovian tax examined in Chapter 4, economic instruments make the polluter worse off; they deliver a potential Pareto improvement but not an actual one. The extra investment arises as the polluters try to avoid the income losses that the control imposes on them.

Compliance costs reconsidered

The final factor bearing on choice of instruments is that of compliance costs. The earlier discussion divided compliance into two elements: monitoring and enforcement. I examine them in turn.

Monitoring

Monitoring costs should be lower for command and control than for economic instruments for two reasons:

1. *It is only necessary to monitor discharges; with economic instruments there are other tasks as well.*
2. *The range of discharges over which monitoring is needed is smaller and the minimum pollution density it is necessary to detect is greater.*

With marketable permits it is necessary additionally to monitor permits, to ensure that the polluters possess the requisite permits for their levels of discharge. With pollution taxes it is not only necessary to monitor pollution but tax liability must be calculated and taxes collected. Subsidies for the installation of pollution-control equipment require monitoring to ensure that the equipment is in fact installed. With subsidies to polluters for pollution reductions there is the calculation and payment of subsidies.

With command and control it is only necessary for the control authority to detect when discharges exceed the consented levels. With taxes and marketable permits, on the other hand, the actual level of discharges has to be recorded in order to determine tax liability or to ensure that the discharges correspond to the permits held. Thus with these economic instruments the entire range of possible discharge rates must be monitored.

The significance of this difference will depend on the nature of the monitoring system. Continuous automatic instrumentation is almost certainly going to be needed with economic instruments, while random sampling may suffice with command and control. For any type of monitoring system, costs are likely to rise as the sensitivity of the system increases. A fairly crude instrument will detect when discharge densities exceed some lower limit, but a refined instrument is needed to provide accurate measures of actual density[1]. Low concentrations may require expensive laboratory analysis but a field kit may be enough to determine whether densities are greater than some pre-specified limit. Command and control requires a less sensitive system and, other things being equal, monitoring should cost less[2].

Enforcement

As noted earlier, all instruments need to be enforced by legal means. While the nature of the offence will differ between instruments there is no reason to suppose that offences will be more frequent for one class of instrument than another and it might therefore be thought reasonable to assume that enforcement costs are the same for all instruments. The argument against this view is that the complexity of the system places differential burdens of fact to be established before the courts. Thus with command and control the facts to be established concern the level of discharge that has occurred. Was it properly measured? Did it come from the defendant's premises? Was the defendant responsible for it[3]? With marketable permits, as well as facts concerning the discharge there are issues concerning the possession of the requisite permits. What if the company had contracted to sell the permits but still had them in its possession at the time, or had contracted to buy them? What if it bought them after receiving notice of the offence but before notice of prosecution? What if it had borrowed permits from someone else, or had lent them? However there may not be more legal issues to be decided since the facts of discharge, with continuous emissions monitoring systems (CEMS), may not be a matter of dispute.

---

[1] Thus with the US Allowance Trading System for SO_2 'measurement of emissions is performed through continuous emissions monitoring systems (CEMS), with quarterly reporting of hourly emissions to EPA (Environmental Protection Agency). The total number of tons of SO_2 emitted by each boiler is then deducted from allowances contained in each electric utility boiler's account, with any excess allowances rolled into the next year's account. If SO_2 emissions exceed the number of allowances held, statutory penalties of $2000 per ton exceeded (indexed to inflation) and an offset of one allowance per excess ton is assessed automatically.' (Rico, 1995, p.117). The capital and running costs of this system are not reported. They are presumably high.

[2] The monitoring system is likely to differ between command and control and economic instruments with some form of CEMS required for the latter. CEMS system may have low running costs but high capital cost.

[3] In the case of discharges to controlled water in England and Wales one question has been whether the water companies, as sewerage undertakers, are responsible for what is discharged into their sewers.

What is likely to be the case is that there are substantial legal costs in switching from one instrument to another since legal rules and precedents have to be established anew.

Classes of instruments compared

The time has come to sum up this extended discussion (see Table 6.1).

We have identified six factors in comparing the costs of alternative instruments: their allocative efficiency; compliance costs; their effects on competition; their impact on innovation in pollution-control technology; their information requirements; and whether they are compatible with the PPP. I consider these factors in reverse order.

All instruments except pollution subsidies are compatible with the PPP. Given that most governments are committed to implementing PPP, pollution subsidies are ruled out for this, among other, reasons.

The tax options and pollution subsidies have high information costs because, to achieve its pollution target, the control authority needs to know MBP curves of the polluters. This information is almost impossible to obtain and for this reason they are usually ruled out as practicable options. Tradable permits are a means of overcoming the information problem of the pollution tax.

All instruments except the pollution subsidy provide an incentive to innovation in pollution-control technology by virtue of the fact that the control imposes costs on the polluter. In addition pollution taxes and marketable permits provide a further incentive to innovation because they impose a financial penalty on pollution that is within permitted limits. The extra innovation in pollution control that they induce may or may not be socially desirable but in any case these instruments are unlikely to deliver the desired amount of innovation. A separate instrument for the purpose would be more efficient. Studies of marketable-permit systems often stress the importance of their contribution to innovation but how far innovation is a result of the instrument chosen and how far a response to new tighter controls on the pollution is debatable.

Barriers to competition are of doubtful relevance to instrument choice. Depending on the precise circumstances, command and control, pollution subsidies and recycled pollution taxes may constitute a barrier to competition.

If information problems rule out tax and subsidy options we are left with two instruments: command and control and marketable permits. The choice between these then involves a trade-off between the efficiency savings of marketable permits and the lower compliance costs of command and control. The efficiency savings of marketable permits have been extensively studied but there is little evidence on differences in compliance costs. Many of the studies of marketable permits as an alternative to command and control either ignore compliance costs, assuming implicitly that polluters obey the law, staying within their consented limits or what is allowed by the permits held, or, where compliance costs are mentioned, assume that they are either trivial or invariant between instruments.

Table 6.1 Factors in instrument choice for control of point pollution

| Instrument | Allocatively efficient? | Compliance costs | | Cost factor | | | |
		Monitoring	Enforcement	Competition barrier?	Innovation incentive?	Information costs	PPP compatible?
Command and control	No	Low	Low	Perhaps	Yes	Low	Yes
Pollution taxes	Yes	High	?High	No	Yes	High	Yes
Pollution subsidy	Yes	High	?High	Perhaps	No	High	No
Recycled tax	Yes	High	?High	Perhaps	Yes	High	Yes
Marketable permit	Yes	High	High	No	Yes	Low	Yes

PPP, polluter-pays principle.

The extent of efficiency savings depends on the spread of the costs of abatement between the polluters. In Fig. 6.4 this is portrayed as the spread of the MBP curves. The greater this spread, the greater the efficiency gains from the use of market instruments. Efficiency savings have in some cases been estimated to be very large. It was estimated that the Allowance Trading System for SO_2 reductions from electricity utilities in the USA would achieve a 40% reduction in annual costs over a command and control system in 1995, rising to 57–65% in the year 2000 and 27–56% in 2010 (Rico, 1995; the increasing range for future dates arises from the uncertainties of predictions). These savings are so large, amounting in 2000 to over $1 billion per annum, that they will almost certainly swamp any differences in compliance costs. However circumstances are particularly favourable to marketable instruments in this case for several reasons:

- The wide range of electricity-generation technologies, embracing not only different fossil fuels from dirty coal to slightly less dirty natural gas but also encompassing non-fossil fuel options of renewables (wind, hydroelectric) and nuclear energy. Utilities can therefore reduce SO_2 emissions by switching between generating plant, expanding supply from less or zero SO_2 polluting sources at the expense of the highly polluting ones.
- The availability of investment in consumer energy conservation as an option. Given pollution constraints, utilities could subsidize consumer investment in energy conservation as an alternative to investing in new generating equipment. This would normally only be profitable if they could raise prices to offset the reduction in electricity demand. The sale of allowances provides an additional source of revenue for this strategy.
- CEMS monitoring is probably needed for SO_2 whatever the control instrument chosen, so the difference in compliance costs between command and control and marketable permits is probably small.

The cost savings are in any event probably overstated because in this industry a sensible command and control system would not fix targets for the individual power plant but for the utility as a whole. This would allow the utilities to utilize the energy conservation option and to realize some of the savings from switching between types of plant, although of course savings from this source are much greater under the Allowance Trading System since they are not limited to the range of plants possessed by the individual utility.

SO_2 emissions from the electricity industry are probably the ideal case for marketable permits and the USA, with a very large number of electricity utilities that limits the risks of thin markets, probably the ideal setting. With other types of pollution the potential efficiency gains are likely to be much less and the difference in compliance costs much greater. Whether marketable permits are to be preferred on cost grounds to command and control is then less clear. Additionally the importance of the location of the emissions is a factor constraining marketable permits as a control option. Location is not an irrelevance in the SO_2 Allowance Trading System but it is not of sufficient importance to rule out the use of the marketable instrument.

Summary

- The choice of instrument to achieve environmental objectives is a major subject of environmental economics and is an important issue for sustainable development.
- This chapter concentrates on the choice of instrument to control point pollution but is applicable to issues other than pollution.
- We consider the principles of choice between broad classes of instruments; in practice technical factors may severely limit or even eliminate choice in specific instances.
- The classes of instruments considered are command and control, and three economic instruments: pollution taxes, marketable permits and pollution subsidies.
- All of these instruments require legal back-up since no instrument is self-policing.
- This means that the control authority incurs compliance costs, costs of monitoring and enforcement, for all instruments.
- Economic instruments face all polluters with the same price of pollution and because of this they are economically efficient in the sense that they eliminate gains from trade.
- Among economic instruments pollution taxes are out of favour because setting tax rates to meet pollution objectives requires information on polluters' costs and revenue functions, which a control authority will not have and can only obtain at excessive cost if at all. Pollution subsidies also have severe information requirements and, in addition, are generally viewed unfavourably as being in conflict with the PPP. Marketable permits avoid the need for this information and are generally advocated by neoclassical economists as their preferred type of instrument.
- Command and control does not eliminate gains from trade. The efficiency effects of economic instruments are the principal advantage claimed for them. It is claimed that they will allow the pollution objectives to be met at lower resource cost than would be the case with command and control.
- However command and control has an offsetting advantage, namely that it entails in general lower monitoring costs and possibly lower enforcement costs than economic instruments. The choice between command and control and marketable permits as the preferred economic instrument depends on a balance between the efficiency gains of economic instruments and the lower compliance costs of command and control.
- Other possible considerations concern whether the instrument presents a barrier to new competition and whether the instrument stimulates innovation in pollution control.
- Boundaries of pollution control typically do not coincide with the boundaries of markets and because of this it is not clear that barriers to competition are an important issue in instrument choice.
- All instruments, including command and control, provide an incentive to the polluter to reduce the costs of controlling pollution. Economic instruments (except possibly pollution subsidies) provide an additional incentive because all pollution, and not simply pollution above target levels, is taxed. Whether this additional incentive is desirable depends on the opportunity costs of investment in new techniques for pollution control. While it is desirable to reduce the costs of meeting environmental objectives it is not clear what priority should be given to it in comparison with other social objectives.

Chapter 7

Nitrate pollution

Non-point pollution

The discussion of instrument choice in Chapter 6 related to problems of point pollution. To operate the instruments considered in that chapter it is necessary for the control authority to be able to identify the polluters and to measure their output of polluting substances. There are many forms of pollution where this is not possible and pollution arising in these cases is known as non-point pollution. The control of non-point pollution is the subject of this and the succeeding chapters. Non-point pollution comes in many varieties and poses a wide range of problems for control that are met with a wide range of solutions. The issues are dealt with by some case studies chosen to illustrate the diversity of problems and solutions. I begin by considering nitrate pollution from agricultural activities, move on to consider household waste and then the problems of exhaust emissions from motor vehicles.

Nitrate pollution from agriculture

The problem

We view this problem from a UK viewpoint but it has parallels in a wide range of other countries. Excessive nitrate concentrations in the environment have social costs in the following ways:

1. Nitrogen applications to old grasslands reduce floristic diversity and some of the rarest and most threatened plant communities are the most vulnerable to degradation from nitrogen applications, e.g. the hay meadows of the North Pennine Dales which are scheduled for special protection under the Habitats Directive of the European Union (EU).
2. Nitrates leaching into surface water courses cause eutrophication with resultant damage to vulnerable and often rare ecosystems with their component plants and animals (e.g. dragonflies).
3. An aspect of this problem is the damage to both commercial and recreational fisheries.

4. Blooms of blue-green algae, for which excessive nitrate concentrations are a necessary condition, are directly toxic to humans and other animals.
5. Where surface waters are sources of drinking water, nitrates are a pollutant with known and suspected deleterious consequences for human health.
6. Permeation of nitrates into groundwater sources render them unsuitable for drinking water supplies and thus impose direct losses on those with abstraction rights to those sources.
7. Ingestion of nitrates directly from green foodstuffs have also been cited as a potential health threat.
8. Finally, volatilization of nitrogen fertilizer as ammonia adds to nitrous oxide levels in the atmosphere and is thus a contributory factor to acid rain.

The significance of some of these pathways is a matter of dispute. Thus the health effects listed at 5 and 7 are questioned. With others, particularly the effects on the aquatic environment, the risks are suspected but not fully understood. Phosphates, rather than nitrates, are the critical factor in eutrophication of freshwater sources but the situation is reversed with coastal pollution. In many cases a monetary valuation of the costs that these pathways impose on society is not technically feasible and, for some of them, is not logically possible. This issue is dealt with fully in Chapter 11.

Nonetheless, the significance of many of these pathways is recognized through environmental standards. Thus the EU imposes a limit on nitrates in drinking water and there are a number of other directives concerned with possible impacts of nitrate pollution on human health. On impacts on the natural environment, the primary protection system in Great Britain is the designation Site of Special Scientific Interest (SSSI) by the statutory bodies for nature conservation in England, Scotland and Wales. In designating an SSSI the nature conservation agencies are required to inform the land-holder of activities that damage the scientific interest of the sites. The land-holder in turn is required to give the conservation agencies notice of intention to carry out any of these damaging activities. Restrictions on nitrogen applications are among the notified damaging activities on grassland SSSIs.

Sources of pollution

Apart from agriculture the major source of nitrate pollution is discharges from sewage works. This, however, is a point source and would be dealt with by the instruments discussed in Chapter 6. Agricultural activities are a major source of nitrate pollution to the environment, through leaching of chemical fertilizers and manure applied to the land, and from leakages and run-off of stored animal waste. Growing crops need nitrogen. In traditional mixed livestock and arable agriculture this was supplied from animal manure either deposited directly on the land by grazing animals or spread by farmers. In modern arable agriculture the nitrogen is supplied largely through chemical fertilizers. Increasing specialization of agricultural enterprises has in the past been a major reason for the growth of nitrate pollution from agriculture. Expansion of the arable area, in so far as it is a separate phenomenon from specialization, has been the other major cause of increasing nitrate releases into the environment. The ploughing up of permanent grassland releases large quantities of nitrates and subsequent use as arable or temporary grass leads to more leaching than is the case with permanent pasture.

The extent to which nitrate use in agriculture should be reduced is open to debate. As explained above, the issue is not simply one of meeting the EU constraints on nitrate in drinking water supplies since nitrates impact on the environment through the other pathways listed above. Determining the appropriate standard is rendered difficult by the scientific uncertainties surrounding the various pathways through which nitrates impact on the environment. The EU directive on nitrate concentrations in drinking water provides a minimum reference point. There is probably not enough evidence of additional environmental risk to justify restrictions beyond this.

Available instruments

As non-point pollution sources, releases of nitrates from agricultural activities pose special problems of control. Since identification of the source is generally impossible[1], discharge consents, pollution taxes and tradable permits are not feasible control instruments. Instead control instruments have to be indirect. There are essentially four options:

1. taxing inputs of nitrogen into agricultural production processes;
2. taxing the outputs of the production process;
3. constraining the production processes used, either:
   (a) in terms of the technology chosen or,
   (b) in terms of the products produced;
4. treating the nitrate once it arrives at its destination in the environment.

Choice of instruments

Option 4 is in a sense a default option, which will happen in any event if no positive measures are taken to deal with the pollution. Nitrate levels in drinking water sources in parts of the UK exceed the EU limits of 50 ppm at some times of the year. In order to conform to the Drinking Water Directive, therefore, the water supply companies who use these sources must reduce nitrate concentrations, which they can do either by installing denitrification plant to remove nitrate from the sources or by blending water from high-nitrate sources with water from sources with lower nitrate concentrations. Where drinking water is derived from underground sources denitrification would in any event be needed after nitrate applications on the water-gathering grounds ceased, since the permeation of water to the aquifers can take many years.

---

[1] This is not the case with leakages from storage tanks for animal wastes and from silage pits, which have been responsible for some major pollution incidents in recent years. These might be considered as point pollution sources but the capacity of farmers to convert them into non-point sources, by spreading their contents on the land, limits the options for control in the case also.

As a solution to the nitrate problem option 4 has three drawbacks: it deals with only one of the pathways listed above, nitrate in drinking water supplies, and does nothing to limit nitrate pollution to the natural environment; it is in breach of the polluter-pays principle (PPP) since the costs fall on the suppliers of drinking water and are ultimately borne by water consumers in higher charges; and it provides no incentives to the agricultural industry to limit existing pollution or to curtail future increases in it. If nitrate pollution is to be reduced along all of the pathways, nitrate emissions must be reduced through one or more of the other options.

There is a further reason why controls on nitrate emissions from agriculture (options 1–3) are preferred to treatment (option 4). Treatment of nitrate has a positive opportunity cost, requiring the installation of expensive denitrification plant or blending works. The opportunity cost of controls on agriculture on the other hand is zero or even negative. The reasons for this are found in agricultural policy.

Under the Common Agricultural Policy (CAP) of the EU there is a gross surplus of agricultural output. Supply exceeds demand at current prices and the surplus cannot be disposed of except at prices that are less than the costs of production. This state of affairs exists because European agriculture is heavily protected, with prices kept high in order to maintain farm incomes. The system is explained in Box 7.1. Under the GATT agreement[2], the EU is committed to reducing agricultural protection, and this can only be achieved by reducing agricultural production to levels that the market will bear. At current productivity levels this in turn requires taking surplus land out of production and reducing productivity on some of the rest. The former is achieved currently through a programme of set-aside, together with restrictions on stocking rates; the latter falls under the ambit of various schemes for environmentally friendly farming and environmental enhancement. This is explained in Box 7.2.

Policies for reducing nitrate pollution work by reducing agricultural output and by inducing switches to less polluting enterprises which, for the most part, are less productive. They thus contribute to meeting the targets for output reduction. Alternative strategies for denitrification involve capital expenditure, which has a positive opportunity cost. Policies for reducing nitrate pollution thus yield a potential Pareto improvement since the gainers from the policy, those who suffer the effects of the pollution, could compensate the losers and still be better off.

[2] General Agreement on Tariffs and Trade. Under this international treaty, which arose as part of the post World War Two settlement, countries are committed to promoting free trade by progressively removing tariffs and other obstacles. GATT works through a series of rounds of agreements on trade liberalization measures. The latest Uruguay Round resulted in agreements by Western countries to reduce agricultural protectionism.

Box 7.2 The CAP

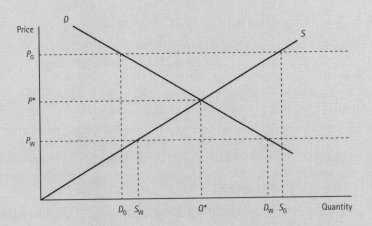

D, S are the demand and supply curves of the EU for some agricultural commodity (e.g. wheat). Supplies of this commodity are available from outside the EU at a world price P_W. At this price EU farmers would be willing to supply a quantity S_W and EU consumers would demand D_W; $S_W < D_W$. The difference D_W-S_W would be imports into the EU from the rest of the world.

The EU authorities decide that, to give a satisfactory income to EU farmers, the commodity should sell at a higher price P_G. To maintain this price, all imports of the commodity into the EU must pay a levy of P_G-P_W per unit quantity. But at P_G EU farmers supply S_G while consumers will only buy the smaller quantity D_G. There is thus a surplus on the market equal to S_G-D_G. To maintain the price P_G the EU authorities must practise 'intervention'; they buy the surplus quantity S_G-D_G at the price P_G. They can only dispose of this intervention stock outside of the EU at the price P_W. There is thus a loss on intervention of $(P_G-P_W) \times (S_G-D_G)$. This is paid by the EU taxpayers and is thus a subsidy to the farmers.

In addition the consumer is subsidizing the farmers as well. The quantity purchased at price P_G would, but for the CAP, have been purchased at P_W. Thus the consumer is subsidizing the farmer by P_G-P_W. But if the price were P_W consumers would have purchased the larger quantity D_W. They have lost therefore an additional consumers' surplus equal to the triangle of height (P_G-P_W) and base (D_W-D_G).

The EU market would clear at a price P^* where demand equals supply. To maintain this price, imports would need to be subjected to a smaller levy P^*-P_W. But EU farmers would be worse off by this policy since they would receive a lower price on quantity Q^* and would not produce and sell the quantity S_G-Q^*. However consumers and taxpayers would be better off.

Box 7.2 Reform of the CAP

The CAP has been under attack for the following reasons:

- The high cost of taxpayer and consumer subsidies.
- The wasteful production of surplus produce, the so-called beef, butter and wheat mountains, and the wine lake.
- Unfair competition: disruption of world agricultural markets brought about by subsidized exports of foodstuffs from intervention.
- Protectionism: exclusion of agricultural imports by the use of import levies.
- Environmental damage: the high and guaranteed prices for EU farmers have encouraged intensification and specialization leading to loss of wildlife habitats, damage to landscapes and archaeological remains, and pollution from the heavy use of farm chemicals.

These attacks have led to the so-called MacSharry reforms. These aim to reduce the costs of the CAP by reducing surpluses (and hence the costs of intervention), and cutting the target prices for intervention crops. Additional measures aim to encourage the spread of environmentally friendly farming. These reforms, however, do not aim to reduce the price to the market clearing level P^* of Box 7.1. *Instead they aim to confine the supply of EU agricultural output to the level that can be sold on the market at the high price P_G.* In the figure of Box 7.1 that supply is D_G. Output reduction is achieved by a policy of set-aside of arable land or restrictions on the number of livestock per area of land, the 'stocking rate' for livestock enterprises. In return for this set-aside farmers receive acreage payments, an annual payment per acre of farmland calculated to compensate them for the income losses sustained as a result of set-aside. Farmers are required to 'set aside' each year 15% of the arable area, rotated about the farm. This set-aside land must not be devoted to production of EU supported crops but must either be left fallow, devoted to cover for game, or used for the production of industrial crops. Thus farmers are being paid not to produce crops. Hence the statement that the opportunity cost of any policy to control nitrate pollution that reduces agricultural output will have zero or negative opportunity cost. It will simply reduce the area that needs to be set aside.

However, options 1 and 2 listed above will reduce farmers' incomes, thus making them worse off from the policy. There will be no actual Pareto improvements unless farmers are compensated for their losses. But if nitrate use is generally reduced in agriculture then, as well as benefiting the environment, there is an added benefit of reductions in the agricultural surplus. Therefore we need to discuss which of the options 1–3 listed above should be used for this purpose.

Apart from the problem of determining reduction targets, there are added complications. The extent of pollution from nitrate use in agriculture varies geographically. It also varies seasonally because the degree of leaching and volatilization are (complex and not fully understood) functions of temperature and rainfall. With regard to leaching to groundwater sources the impact depends also on water volume and flow rates. Pollution also varies with the ground cover, being highest from bare ground and when the planted crop is quiescent.

These considerations argue for the value of a policy under option 3(a), namely a scheme to encourage environmentally sound practice with regard to fertilizer use. Such a scheme could only be voluntary since a code of practice could not feasibly be enforced except at unacceptably high cost. The UK Ministry of Agriculture, Fisheries and Food (MAFF) publishes and recommends a code of good agricultural practice for fertilizer use. One might suppose that this code would be self-enforcing since it is designed to ensure that fertilizer is not wasted but is applied in the quantities and with the timing to maximize the farmer's returns from fertilizer use. But environmentally sound practice is not necessarily the same thing as good agricultural practice, which is designed to ensure the maximum efficiency from the use of agricultural inputs. While one might suppose in general that fertilizer applications that leach away must be inefficient, this is not necessarily so. Some losses from leaching may be deemed acceptable in the interests of higher productivity. But, to repeat the point made previously, there is no social benefit from increasing, still less from maximizing, agricultural productivity, since any increases in output achieved by this means simply increases the land that has to be set aside.

Environmental damage from fertilizer use in excess of the environment's absorptive capacity appears acceptable to the farmer because she does not pay the costs of the consequent environmental damage. Thus some additional instrument is needed to reconcile environmental and agricultural optimum use rates of fertilizer. Codes of practice are unlikely to work on their own.

We have so far ruled out option 4 and have found that option 3(a) on its own would not be sufficient to deal with the problem. I now consider the remaining options.

Taxation options

A sales tax on nitrogenous fertilizer would be based on the nitrate content of the product. Such a tax would conform to the PPP. It would be economically efficient since farmers would face the same tax and would adjust their nitrate use to the point where, at the margin, the benefit from additional nitrate use equals the tax. However it would be inefficient environmentally because there is no unique mapping from nitrate input to nitrate output. The same quantity applied at different times in different places to different crops would result in different degrees of environmental pollution.

A sales tax on agricultural output would be scaled by some calculation, necessarily crude, of the relative contributions of different commodities to the problem of nitrate pollution. A tax on agricultural output would also conform to the PPP and would have the same efficiency properties in terms of economics since all farmers would face the same tax. It would, however, be environmentally more efficient since it would correct for differences in the polluting properties of different crops. It would, of course, do nothing to address the other sources of variation such as the timing of fertilizer application.

Both instruments would have a positive effect on the agricultural efficiency of nitrate use, since the opportunity cost of overuse would be increased. Provided that there is some positive mapping between nitrate input to the land and nitrate leaching from it, and this is open to some doubt, both forms of taxes would also have some positive effect on reducing the damage from other sources of variation (timing, location) although, because neither is targeted to these problems, they would not be efficient devices in those regards.

Chemical fertilizer is not the only available source of nitrate for agriculture; there is also farmyard manure, processed sewage sludge and other organic wastes. In general these are cheaper but less efficient sources of nitrogen. The input tax would have the effect of raising the price of efficient chemical fertilizers relative to less efficient manure and would encourage the substitution of the latter for the former and would therefore require the parallel use of some other instrument to control manure use. The output tax has the advantage that its impact is independent of the source of nitrate input so that it would provide no price incentive for the substitution of manure and other nitrate sources for chemical fertilizers.

This conclusion needs qualifying. If manure is a cheaper but inferior input for agricultural production then its use could rise with either taxation instrument because of an *output effect*, i.e. the effect of the instrument in reducing the volume of agricultural output. However the substitution would be greater with the fertilizer tax because a *relative price effect* (nitrogen from manure has become cheaper compared with nitrogen from chemical fertilizer) would be reinforcing the output effect. This is explained in Box 7.3.

If manure applications carry a greater risk of excess phosphate pollution of the environment, which, as already noted, is also a cause of environmental concern, then this is an additional argument in favour of the output tax option.

An output tax would work in a free competitive market for agricultural produce but that is far from the current reality. The case against an output tax obviously hinges on the complexities of agricultural price support in a highly managed market. The CAP relies principally on land set-aside (see Box 7.2) to bring the market closer to balance and deals with environmental problems by special schemes accompanied by compensatory payment. In this system the structure of output prices bears no necessary, and probably no actual, relationship to that which would prevail in a free market. Adjustments to these artificial prices to control nitrate emissions would be administratively difficult (or politically difficult or, realistically, both) and could well have perverse results.

Thus a tax on artificial nitrogenous fertilizer is probably a better practical option for the control of nitrate emissions from agriculture. For reasons given above, it would probably need to be accompanied by controls on the use and disposal of animal wastes since taxation of manure production is not practicable.

Box 7.3 Output and relative price effects of instruments to control nitrate pollution

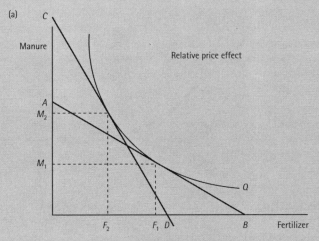

(a)

Q is a contour on a production function for an agricultural product that uses both artificial fertilizer and manure as inputs. The initial prices of the inputs are given by the slope of the line AB and quantities F_1 and M_1 are used. A nitrate tax raises the price of artificial fertilizer and the post-tax relative prices are given by the slope of the line CD. Artificial fertilizer use falls to F_2 but manure is substituted for it, rising to M_2.

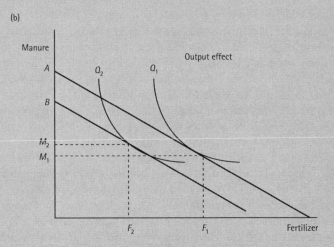

(b)

An output tax has reduced the output of the agricultural commodity from Q_1 to Q_2. Since it has not changed the relative prices of the inputs the price lines A and B have the same slope. The reduction in output leads to a fall in the use of artificial fertilizer from F_1 to F_2, but because a different technique of production is chosen, the use of manure has increased from M_1 to M_2.

Box 7.4 Elasticity of demand

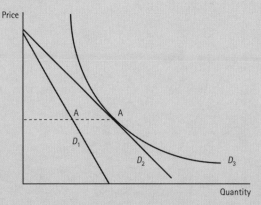

Elasticity of demand is defined as the ratio of the percentage change in quantity demanded to the percentage change in price. Thus if the quantity demanded of a product or factor falls by 2% when price increases by 1% the elasticity is −2. If the elasticity is less than −1 demand is inelastic; if the demand is greater than −1 the demand is said to be elastic. If demand is inelastic then total expenditure on the product or factor rises when price goes up. Conversely when demand is elastic total expenditure falls when price rises. For a straight line demand curve such as D_1 the elasticity declines as one moves down the line. This is because the slope of the line is constant but as one moves down it, to the left the price decreases so that a given absolute change in price becomes a larger and larger percentage change; the reverse happens with the quantity demanded. Since they start from the same point and D_2 has a shallower slope than D_1, D_2 is more elastic at every price than D_1. Point A is the point where elasticity is equal to −1. To the right of A demand on D_1 and D_2 is inelastic; to the left of A it is elastic. To have a constant elasticity a demand curve would have to have a varying slope. D_3 is a constant elasticity curve. The minus sign is conventionally ignored in discussions of elasticities.

The normal argument adduced against a fertilizer tax is that the demand for fertilizer is highly inelastic (Box 7.4) so that large falls in farmers' incomes are required to achieve relatively modest falls in fertilizer applications. Inelastic demand for a factor exists when there is little opportunity, by change of technique, to substitute other factors for it when its price changes. Production requires a certain quantity of the input; other inputs perform different functions and are not substitutable; there are no alternative inputs available but not currently used that can act as substitutes. This is often the case for nonrecyclable basic raw materials.

The view that demand for artificial fertilizer is inelastic rests on a number of studies of the response of agriculture to changes in fertilizer prices based on economic models of the behaviour of agriculture in a number of Western countries.

These studies typically concern short-term elasticities, i.e. they relate to the immediate response of farmers who have invested in machinery which locks them in to specific techniques and to a narrow range of product choice. Elasticities in the long term should theoretically be much higher.

If reducing nitrate pollution via nitrate tax does look likely to cause unacceptable losses of farm incomes, it is open to the EU to adjust its considerable package of farm subsidies to offset the income effects while retaining the reductions in nitrate pollution.

Restrictions on land use

The current approach in the UK to controlling nitrate pollution from agriculture utilizes the final option that we have listed, option 3(a): constraining the production processes used. This is done in two contexts and for two purposes:

1. in areas where nitrate leaching is deemed to be particularly damaging because it endangers the UK's ability to meet the requirements of the EU Drinking Water Directive;
2. in areas where nitrate restrictions are deemed necessary to protect valuable wildlife habitats.

The former are the catchment zones for sources of drinking water, which are designated as Nitrate Sensitive Areas. Reductions in nitrate leaching in these areas are to be achieved by agreement with the farmers who accept restrictions on land use in return for compensation payments based on the size of their land holdings and designed to provide compensation for the loss of profits resulting from the restrictions. The scheme is presently voluntary but with the threat of compulsory compliance in the wings should the voluntary approach fail.

The latter are management agreements to protect specific wildlife habitats in SSSIs and in Environmentally Sensitive Areas (ESAs). Under SSSI agreements the restrictions on land use are tailored to the specific requirements of the sites; with the larger ESAs common conditions apply to all participants within a specific ESA. Compensation in both cases is again related to profit lost as a result of the restrictions.

This approach obviously deals with only two of the pathways listed at the beginning of this chapter and ignores the issue of nitrate pollution of the wider environment. It is also in conflict with the PPP.

ESA payments and management agreements for SSSIs relate to other restrictions as well as fertilizer applications, such as pesticide use, stocking rates and maintenance of stone walls and field barns. The Nitrate Sensitive Areas scheme, being of more recent invention, is presumably modelled on the ESA scheme which is also administered by MAFF. The SSSI scheme is administered by nature conservation agencies answerable to a different government department, the Department of the Environment. It is sanctioned under the Wildlife and Countryside Act 1980 and not partially funded from the European agricultural budget.

There are, however, some obvious differences of principle between nitrate pollution and these other environmental protection measures. ESA and SSSI payments are designed to protect wildlife and landscape by maintaining agricultural practices that would otherwise be abandoned. In some cases farmers are being paid to maintain hay meadows when it would be more profitable to cut grass for silage, or to maintain stone walls where wire fences and larger fields are cheaper, or to keep grassland where the land would be more profitably put under crops. The theory of public goods may be invoked here. In farming in traditional ways the farmer is conferring a benefit on society in terms of maintaining rare plants and animals or providing a pleasant landscape. She is unable to capture these benefits since the goods concerned are non-excludable. In the absence of these payments, other less environmentally acceptable practices would be followed or, if these were prevented by some form of environmental standard backed by sanctions, the land would be abandoned.

Nitrate pollution is not like these cases; it is simply a negative externality. The farmer is not being asked to undertake any positive management on behalf of the environment; she is being asked to refrain from actions that damage it. This is a classic case of pollution to which the PPP should apply.

The case for payments to the farmer in return for restrictions that reduce nitrate pollution is that they apply only to some farmers who happen to be living in water catchment zones. These farmers are therefore disadvantaged in comparison with other farmers. There is no intention to apply the restrictions generally and indeed it would not be practicable to do so, since the compliance costs would be prohibitive. The scheme is only practicable *because* it is confined to restricted areas (as of course is the case with ESA payments and SSSI management agreements). Nitrate pollution is therefore locationally specific: the same emissions of nitrates emitted in other areas would do less damage. In any event there can be no guarantee that a nitrate tax levied on the sales of chemical fertilizers would produce sufficient reduction in nitrate pollution in the Nitrate Sensitive Areas. Thus there would still be a need for additional measures for these areas.

Summary

- Pollution control, as discussed in Chapter 6, requires that the control authority can locate the sources of pollution and identify the polluters.
- Where this is not possible we have a problem of non-point pollution. With non-point pollution the instruments discussed in Chapter 6 are not appropriate.
- Nitrate pollution from agricultural activities is a classic example of non-point pollution. Nitrogenous fertilizers are widely used on agricultural land and pollute water sources and the atmosphere. It is not possible to identify all the polluters or to attribute particular polluting flows to particular sources.
- Nitrate pollution affects water sources and damages aquatic ecosystems. In high concentrations it is also a problem in drinking water. It can also be a threat to rare plants when applied to old grasslands and can contribute to acid rain.
- The reduction of nitrogen use by agriculture will result in improvements in all of these problems.

- However drinking water problems can also be tackled by treatment of the polluted water. With underground sources, denitrification would continue to be necessary for extended periods after nitrogen applications are reduced because of the length of time taken for surface water to permeate to aquifers.
- Apart from use as a temporary measure for underground sources, denitrification is not a cost-effective option for drinking water since it absorbs resources that have alternative uses. The opportunity cost of reducing agricultural production, and with it nitrate pollution, is zero since agricultural produce is in surplus within the EU and land is being taken out of production.
- Possible economic instruments are a tax on nitrogen inputs and a tax on agricultural output, graded by the contribution of the product to nitrate pollution.
- The distortion and manipulation of prices within the CAP makes the output tax infeasible. An input tax is therefore the best option.
- The alternative is physical controls on agriculture in water catchments. A scheme, the Nitrate Sensitive Areas scheme, has been introduced in England and Wales under which farmers receive payment in return for accepting restrictions on agricultural operations.
- Payment is justified because the restrictions only apply to some farmers who are thereby disadvantaged compared with others.

Chapter 8

Domestic waste

The problem

The problem of domestic waste probably needs little explanation. Households, in Western countries at least, generate it in considerable quantities. In the UK in the early 1990s households were generating approximately 17 million tons of domestic waste per annum (data in this section are taken from Environmental Resources Ltd, 1992). If left to individual disposal this waste would pose major problems of public health and pollution as well as seriously damaging civic amenity. In consequence it is collected and disposed of, usually by or though local government. Individual disposal, except under controlled conditions, is illegal, although 'fly tipping' remains a problem and is a potential constraint on solutions to the domestic waste problem. In the UK, 13 million tons per annum is collected from households, with the other 4 million tons being deposited by households either at civic amenity sites or at recycling collection points (bottle banks, etc.). Collected domestic waste is principally disposed of by landfill, which accounts for 90% by weight; 9% is incinerated and about 1% disposed of in other ways including recycling. The principal environmental problems are thus largely connected with landfill.

First, there is the amount of land that is absorbed. Traditionally landfill occurred on land of low agricultural value and waste tipping was seen as a strategy for land reclamation. Thus a number of coastal authorities located their tips on coastal marshes and saltings and saw tipping as a means of adding to the supply of economically useful land. Inland, favoured sites were worked-out gravel and sand pits, peat bogs and other areas of low or zero agricultural value. The choice of these sites was partly economic since land values on them were low. However these sites were often important ecologically and concern with wildlife conservation has presented waste-disposal authorities with a problem in the form of a shortage of suitable sites for landfill. Landfill can no longer be presented as yielding a community benefit and any proposed new landfill site faces public opposition.

Amenity problems of landfill are the build-up of methane from the decay of organic matter and pollution of water courses from run-off. Additionally there are problems of smell, rats and visual disturbance from open sites and contamination of the soil of filled sites with residues of heavy metals and chemicals. The alternative of incineration carries problems of air pollution.

The final environmental issue concerns the waste of raw materials involved in burying waste that could be *economically* recycled.

Solutions

If the problem of domestic waste is seen to lie in the environmental costs of landfill there are two broad classes of solutions to it: to reduce waste production or to recycle it. These solutions are of course not mutually exclusive and the optimum solution may be some combination of reduction in waste creation and the recycling of what remains. However if the problem is seen as primarily that of the waste of raw materials then recycling is predicated as the only solution, regardless of the volume of waste produced. Official perceptions of the problem centre around recycling. The consultation draft of the UK Government's Waste Strategy for England and Wales (Department of the Environment and Welsh Office, 1995) suggests targets to stabilize household waste production at the 1995 level, recycle 25% of the domestic waste stream by the year 2000, and reduce the proportion of waste going to landfill by 10% over 10 years. The EU Packing and Packaging Waste Directive sets recycling targets of 50–65% for packing waste by 2001. Since it is at the centre of the strategy I start by considering recycling, returning to the issue of the alternative solution of reducing waste production at the end of the chapter.

Recycling of household waste might be justified for the following reasons:

1. to reduce the demand for landfill and the problems associated with landfill sites as listed above;
2. to save exhaustible resources, such as metals, and thereby increase the life of existing known stocks;
3. to avoid the environmental damage associated with the extraction and processing of primary materials.

For the first reason, a waste-disposal authority faced with problems of finding sites for landfill might be moved to adopt a recycling strategy on its own volition. However it could be argued that the incentives for it to do so are inadequate because the authority does not bear all the external costs of its landfill strategy. There is, in other words, a problem of externalities which are not internalized to the decision-maker.

The second reason suggests another source and form of market failure. If recycling is not taking place to save exhaustible materials then, provided that the recycled materials are adequate substitutes for primary materials, this means that the market price of primary materials does not reflect their future scarcity and in consequence does not provide a sufficient incentive to recycle. The economics of exhaustible resources is discussed in Chapter 13.

The third reason relates to another source of market failure in the market for primary materials. This is that the externalities of material extraction and processing are not internalized to the extractors so that the prices of primary materials do not reflect their social costs.

In all three cases the issue hinges on the economics of recycling since it may be that recycling consumes more resources than it saves.

Economics of domestic waste recycling

Table 8.1 gives some data on the domestic waste stream in the UK and the possibilities of recycling. These data refer to waste collected from domestic premises. They exclude commercial waste, amounting in the UK to some 3 million tons per annum, which has typically much higher recycling prospects, and the material delivered by households to what are called 'bring' facilities (bottle banks, etc.) most of which is recycled as intended. The recycling ratios given are technical maxima. They are less than 100% because recycling is limited by contamination of the material and because of bonding and mixing of materials during manufacture (metal/metal, metal/plastic and mixing of plastics). Practical recycling ratios will be considerably lower than these technical maxima for a number of reasons including constraints imposed by the initial means of collection.

The economic limits on recycling are dominated by two factors: the costs of collection of the material for recycling, and the markets for the resulting recycled materials. Collection costs are high so that even if recycled materials are of comparable quality to primary materials cost conditions limit the market for them. Primary materials are typically produced at one of a small number of sites and are delivered and processed in bulk. With recycling from households[1] small quantities are collected from each of thousands of locations and, even if given a primary sorting by the household, have to be taken to a central point for additional sorting before going to the recycling plant.

The economics of recycling are greatly improved by maximizing the use of 'bring' operations and getting households to pre-sort collected, or as it is usually

Table 8.1 Composition of domestic waste and recycling possibilities

Material	Percentage of total waste (by weight)	Percentage of material recyclable (technical limit)
Paper and board	30.2	59
Plastics	8.4	55
Glass	9.6	90
Ferrous metals	7.0	79
Aluminium	0.01	75
Textiles	1.9	52
Compostables	28.2	50
Other	14.1	0
Total	100	52–63

Source: Environmental Resources Ltd, 1992, Table 3.9.

[1] This may not be true with some forms of commercial waste.

called, kerbside waste. There appears to be little difficulty in persuading households to practise some sorting since the enthusiasm for recycling is high. But the sophistication that can be attained by domestic sorting is low and may be inadequate for effective recycling. This is the case with plastics, where sorting by chemical type is necessary for effective recycling but is deemed not to be possible by households. It is also the case with grades and types of paper beyond a division between newspapers and other 'mixed' paper. The most successful waste streams for domestic sorting are glass, where the principal sort required is by colour, and aluminium and steel cans, where a magnetic sort at the 'bring' facility suffices. Commercial bottle banks and aluminium container banks are the most successful 'bring' facilities. Sophistication of household sorting limits the development of 'bring' facilities beyond these products. Even were more sophisticated sorting by households possible the small quantities generated of each type per household would render kerbside collection uneconomic.

There are basically two systems of kerbside collection of sorted waste: the so-called 'blue box' system where households leave their dry recyclable wastes at the kerbside and sorters accompany the collection vehicle; and wheeled bin systems where households are given wheeled bins into which they deposit sorted waste. The degree of sorting under the wheeled bin system depends on

Table 8.2 Kerbside collection costs for different systems*

System	Collection frequency	Recycling potential	Cost/household (£/annum)†	Index of costs
1 One wheeled bin unsplit	Weekly	Zero	24.4	100
2 One wheeled bin split	Weekly	Low (target of 25% domestic waste recycled not met)	29.3	120
3 Two wheeled bins unsplit	Twice weekly	Moderate (target of 25% domestic waste recycled not met)	29.9	123
4 Two wheeled bins, one split	Twice weekly	High (allows 25% target to be met)	34.1	140
5 Blue box	Twice weekly‡	High for dry recyclables but will not meet 25% target unless separate collection of organics is added	38.0	156

* All options assume the existence of 'bring' facilities where glass and aluminium cans are collected.
† 1992 prices.
‡ One collection of unsorted waste; one blue box collection.
Source: Leeds City Council, Draft Recycling Plan, July 1992 to June 1993.

the number of bins provided and whether the bins are compartmentalized. The practical maximum is seen as two bins and the maximum division in a bin to be two also, so that the maximum household sort is a fourfold one. Since one of the categories of waste will be residual 'trash' this gives a maximum threefold sort for recycling. Systems may be supplemented, for those with gardens or allotments, with the provision or sale of household composters, thus reducing the collection and disposal of organic waste.

Collection for recycling requires capital investment in different collection vehicles as well as investment in wheeled bins. Costs rise substantially with the degree of sort. Thus a study by Leeds City Council of the options gave the differences in collection costs shown in Table 8.2. This suggests that even a moderate amount of kerbside recycling carries a substantial financial penalty and more elaborate schemes, which allow the target of 25% of household waste to be recycled, increase collection costs by 40% and upwards. The blue box system is substantially more expensive than wheeled bins.

But collection costs are only a part of the additional costs of a recycling strategy for domestic waste. The household sort is only a preliminary one and any recycling scheme has to allow for additional sorting and processing, as well as packaging and despatch at the collection depot. If organic waste is presorted in the collection system provision has to be made also for a composting plant. The Leeds City Council study found a depot cost for dry recyclables of £3 per household and a composting cost of £0.85 per household. The addition of these items raises the costs of kerbside recycling to 32% above the no-recycling system for the cheapest unsplit-bin system (system 1 in Table 8.2), which would not have composting of organics (only dry recyclables could be sorted), and to 55 and 71% above no recycling for the high recycling options 4 and 5.

Against these considerable increases in the cost of household waste collection and disposal have to be set two sources of potential revenue: the sale of the recycled materials, and the savings in landfill costs of disposal.

The value of the recycled material, once the most readily marketable materials have been creamed off by 'bring' systems, is in fact very low. In the Leeds study over 70% of the recovered material was paper; 60% of this was newspaper for which the market was saturated. It could at the time be sold for a low price but it was thought that the price would fall to zero in the near future. There was no market either for the rest of the paper except that small quantities of cardboard might be recovered by additional sorting at a considerable extra cost. A small amount of revenue was obtainable from sorting plastics and from metal cans that had not been taken to 'bring' facilities. But overall the revenue generated from the sale of recycled materials did not cover the depot costs and made no contribution at all to collection costs. This conclusion, that recycling of household waste is a loss-making activity, is reached also in the Environmental Resources Ltd (1992) study referred to earlier.

The conclusion to be drawn is that if recycling is to be a worthwhile investment for the disposal of household waste then it can only be so by virtue of the savings in landfill costs that follow from a reduction in the quantity of waste to be disposed of. Both the Leeds and the Environmental Resources Ltd studies found that landfill savings at current and forecast prices of land would not be sufficient to make the recycling effort profitable. *A fortiori* the more expensive schemes, which allowed the authority to meet the target of 25% household waste recycled, would be unprofitable and would involve households in a considerable increase in the costs of waste disposal as well as, through the responsibilities for pre-sorting, increasing their input into the process.

Instruments for the recycling of domestic waste

The discussion of the economics of recycling domestic waste suggests that waste-disposal authorities will not have an economic incentive to recycle kerbside waste. However, recycling may still be justified for any or all of the three reasons suggested above, namely to avoid the externalities of landfill and of primary material production and to extend the life of stocks of exhaustible resources. The externalities are mostly forms of pollution which, for reasons given in Chapter 5, are probably not open to monetary valuation. The pollution control authority might nonetheless seek to impose standards and the UK Government and EU targets for waste may be viewed as such standards. The issue in this section, therefore, is what sort of control instruments should be used to achieve an increase in recycling of kerbside domestic waste.

Figure 8.1 depicts a simplified version of the waste stream. In order to focus the discussion we assume that recycling and landfill are the only disposal options. Thus incineration, dumping at sea and sending waste abroad for processing and disposal are ignored. All of these other possibilities carry substantial, if hidden, environmental costs.

In the discussion of the economics of recycling we assumed, implicitly, that collection and disposal were integrated into one body. That of course need not be the case; collection and disposal are separate operations and it is useful here to assume that they are conducted by different bodies. The first question that has to be settled in any discussion of instruments is where in the waste stream the instruments should be directed. A simple model of the process allows us to concentrate on instruments directed at the point of disposal.

Assume that there is one disposal authority and one collection authority for each area (city or local government division). The disposal authority covers its costs by charges on the collection authority and the collection authority covers its costs in turn by collection charges on households. Households are not paid for direct recycling and of course they cover their private costs of sorting and delivery to 'bring' facilities. The objective of the pollution control authority is to reduce the volume of disposal to landfill and to redirect it to recycling. I consider options of economic instruments and command and control.

The control authority might consider the following options, all of which would be imposed on the disposal authority:

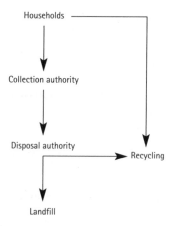

Fig. 8.1 Domestic waste flows.

1. Economic instruments:
   (a)   a tax on landfill;
   (b)   transferable (marketable) landfill permits;
   (c)   a subsidy on recycling.
2. Command and control:
   (a)   controls on the amount of waste permitted to go to landfill;
   (b)   a requirement that a minimum proportion of delivered waste should
         go to recycling.

The UK Government has in fact chosen a combination of three of these. It has
introduced a landfill tax and recycling credits and has set a target for the minimum
proportion of waste going to recycling. I consider these instruments in turn.

Landfill tax

The landfill tax would probably be levied on the weight of waste going to
landfill. Faced with a tax on landfill the disposal authority will pass it back to
the collection authority in the form of higher disposal charges. Provided it is
able to do this, it has no incentive to substitute recycling for landfill. The collection authority will pass the charge backwards also in the form of higher
collection charges on households.

What happens then depends on the nature of the charging system practised by
the collection authority. The typical system is for the collection authority to levy
a fixed annual charge on households, not differentiated by the volume of waste
the household produces and set so as to cover the total costs of collection and disposal. With this system households have no incentive to reduce the volume of
waste that they generate for disposal since they receive no financial benefit from
doing so. The financial savings that result from any effort they make in reducing
the volume of waste will be spread across all households and, since the number of
households is likely to be very large, will be dissipated to the point of insignifi-

cance. Furthermore any savings could easily be negated by increased waste from other households. This charging system thus is said not to be *incentive compatible*. An incentive-compatible system would be one where households are charged according to the volume of waste they produce for collection. An example of such a system would be one where waste is only collected in official disposal bags that households have to buy from the collection authority. In this case a rise in the costs of collection would appear as a rise in the price of disposal bags. Households would then have an incentive to reduce the volume of waste for disposal. They might do this is in a number of ways:

1. by increasing the amount of waste delivered to 'bring' recycling facilities;
2. by reducing their volume of waste-intensive consumption;
3. by self-disposal of waste, e.g. by:
   (a) composting organic waste;
   (b) burning combustibles;
   (c) putting waste in public litter facilities and other households' bags;
   (d fly tipping.

While two of these (1 and 3a) contribute to the recycling objective and 2 is obviously desirable, the other responses are both unhelpful and undesirable. In summary, without incentive-compatible charging of households a landfill tax will not work. With incentive-compatible charging the tax will have some effect of reducing the quantity of waste going to landfill but the stream will not necessarily be either reduced in volume or diverted to recycling.

Transferable landfill permits

The principles of transferable permits should now be familiar. The control authority decides how much landfill to permit and produces permits for that amount. It can either 'grandfather' them to disposal authorities or auction them. Disposal authorities are only permitted to landfill to the extent of the permits they hold and they may trade in permits with other disposal authorities. The control authority will need to monitor the system and enforce it. Given that it does so, the price of the permits will settle at the level at which disposal authorities are collectively choosing the amount of landfill that the control authority has set.

The permits are an income-earning asset to the disposal authorities and thus provide them with an incentive either to recycle waste or to reduce the amount they receive for disposal. To do this they have to trade with the collection authorities to persuade them either to reduce the quantity of waste delivered or to deliver it sorted for recycling. Thus the financial savings from selling permits will be captured in part by the collection authority.

To deliver sorted waste the collection authority must go through the processes described in the section on the economics of recycling. We saw there that households are willing to sort their waste if given the facilities (wheeled bins, etc.) to do so. The collection authority can reduce waste volume by giving or selling composting systems to households. It might also refuse to collect waste (e.g. bottles that are suitable for 'bring' recycling) but this is difficult to enforce. It might again attempt to provide households with a financial incentive to reduce their waste, for instance by levying an extra charge for collections exceeding a certain volume per household but this will result in households resorting to antisocial activities like fly tipping.

We saw that kerbside recycling is generally an unprofitable activity. Tradable landfill permits work by making the cost of landfill rise to the point where (leaving aside minor effects on reductions in household waste production) recycling appears to be profitable to the disposal and collection authorities. One is led to ask who pays for the losses on recycling that appear in this system as costs of landfill permits. The answer is clear: the costs will ultimately fall on the households in whatever form they pay for waste collection and disposal.

We saw how waste collection charges that households pay would rise if recycling were adopted. Tradable permits would only result in smaller rises than these if there were efficiency gains to be realized in recycling by concentrating it in some disposal authorities and thus realizing economies of scale (Box 8.1) If economies of scale exist then some disposal authorities would sell their landfill permits and concentrate on recycling; others would buy permits and concentrate on landfill.

It is not clear whether economies of scale do exist in recycling to a sufficient extent to allow specialization between disposal authorities. Even if they do whether they should be realized by encouraging specialization depends on the nature of the problem that the policy is designed to alleviate. Thus if society wishes to discourage landfill and encourage recycling to avoid the externalities of primary material production, to extend the life of exhaustible resources, or to reduce emissions of methane, a powerful greenhouse gas, into the atmosphere, then the efficiency gains reduce the costs of meeting the objective. On the other hand if the problems are site specific, that landfill creates local pollution and damages local amenity, perhaps destroys valuable wildlife sites, then the distribution of landfill that would result from marketable permits may be unacceptable.

Recycling subsidies

Since the main reason why disposal authorities resort to landfill is that recycling kerbside domestic waste is a loss-making activity, a subsidy on recycling might be thought to be a suitable instrument. The subsidy would presumably be proportional to the weight of waste recycled. Monitoring by the control authority would of course be required. If paid to the disposal authority some of the subsidy would have to be transferred to the collection authority to cover the costs of providing sort facilities for households. The principal difference that this instrument provides over the others so far considered is that it would not result in increases in waste collection charges to households since the recycling losses are paid for out of national taxation. This might be appropriate in so far as the reasons for encouraging recycling are not local amenity. In addition national taxes are more progressive than local taxation. Households typically pay the same waste disposal charge regardless of size and income level.

Box 8.1 Economies of scale

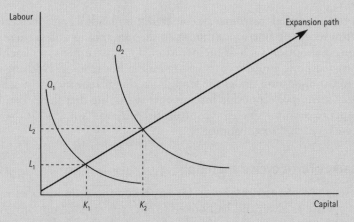

The diagram shows two contours of a production function. Along the expansion path the factors of production are combined in a constant proportion L/K. There are obviously many expansion paths (any straight line through the origin) each representing a different factor proportion L/K. If the relative prices of factors are constant, the increase in output will involve points on an expansion path. In shifting from contour Q_1 to Q_2, factor input increases in the proportion L_2/L_1 ($=K_2/K_1$). If output increases in a greater proportion than factor input, i.e. $Q_2/Q_1 > L_2/L_1$, the production function is showing increasing returns to scale. In this case, provided factor prices are constant, an increase in output will result in falling unit production costs. For this reason increasing returns to scale are otherwise referred to as economies of scale. If output increases in a smaller proportion to input ($Q_2/Q_1 < L_2/L_1$) then the production function will exhibit decreasing returns to scale and expansion will face rising unit costs. Constant returns to scale mean that $Q_2/Q_1 = L_2/L_1$ and unit costs are constant.

Unlike the landfill tax, a recycling subsidy, if set at a level sufficient to offset the losses from recycling, would achieve its objective but contains no incentives to efficiency.

Command and control

The economic instruments that we have discussed are appropriately directed at the disposal authority, which will pass the incentives back to the collection authority. The alternative to economic instruments is regulation. Regulations about the quantity of waste going to landfill or to recycling can only be imposed on the control authority in parallel with regulations governing the behaviour of the collection authority. The appropriate regulations within the framework of our simple model of the process would require the disposal authority to accept all waste delivered to it by the collection authority but give it the give the power to specify that it must be subject to the necessary household sort.

The two suggested instruments, constraints on landfill and specification of a minimum recycling fraction, amount to the same thing unless there is some third disposal option such as incineration. The instruments will achieve the necessary recycling and the costs will fall upon households. Nothing need be added to the discussion under the landfill tax. Regulations will not achieve the efficiency savings of economic instruments, assuming that they exist, but will be cheaper to monitor since the control authority will know what the requirements are on each disposal authority.

The market for recycled materials

The discussion of instruments so far has taken the economics of recycling domestic waste as given and has looked to alter the disposal options. An alternative approach is to seek to raise the demand for the products of recycling domestic waste. Instruments that might be used here include the following:

1. taxation of primary materials;
2. subsidy to the processors of recycled materials;
3. regulations governing the quantities of recycled materials to be used in manufacture.

How a tax on primary materials would operate would depend on the nature of the product. The simplest case would be with metals; a steel mill or an aluminium smelter could be taxed on the quantity of ore that it used. Since all metal ores have to be processed the tax could be levied at the mine, or, for imported ores, at the port. Glass and plastic manufacture use no material that is unique to them (no glass or plastic ore!) and the tax therefore would have to be levied on the finished product. Thus for glass the tax would have to be on glass containers graduated by the amount of recycled glass used in their manufacture. This would be similar with plastic containers. Paper manufacturers would face a tax related to the quantity of recycled paper used in manufacture.

The incidence of such taxes would depend on the elasticity of demand for the products. If the demand is highly elastic the producer would not be able to pass the tax on to her customers but instead would be forced to absorb it herself. The higher the elasticity the greater the proportion of the tax absorbed by the primary producer and the greater her incentive to increase the use of recycled material to avoid loss of profit.

The principal objection to such taxes is that in very many cases they would be difficult to monitor and enforce since there are no simple means of determining the quantity of recycled material that a product contains. The producer would therefore have an incentive to overstate the quantity of recycled material and the taxing authority would have to prove her wrong. This is not a problem with metals, where the tax could be levied on the delivery of ores. The problem of monitoring and enforcement applies equally to the third policy option of regulating on the proportion of recycled material used.

The disposal authority delivers recycled material to the producer. A subsidy on the delivered material would encourage the producer to substitute it for primary material. The subsidy would be easier to monitor than the tax on primary material input since it would be payable on quantities delivered from disposal authorities. Since, as we have seen, recycling of domestic waste poses real costs on collection and disposal authorities, the disposal authority would have to capture some of the subsidy in the form of higher prices paid for recycled materials, and to transfer some of this recaptured subsidy to the collection authority to pay the additional costs of household sorting. This instrument is thus closely allied, indeed an alternative form of the recycling subsidies considered above. It also places the costs of the system on the taxpayer, rather than the households as producers of waste, and provides no incentives for efficiency gains.

Waste reduction

The instruments that we have so far considered provide, at best, poor incentives to households for waste reduction. Incentive-compatible charges on households convey some incentives in this direction, but as we saw they can also lead households into antisocial actions which certainly do not help in the solution to the household waste problem. Obviously, if it is possible to achieve it, reducing waste production is desirable. Whatever disposal method is used, the costs of disposal will decline with the quantity of waste that has to be dealt with. Both landfill and recycling absorb real resources and carry negative externalities.

Table 8.3 presents an alternative classification of the household waste stream. Discussion of the possibilities of reducing the volume of waste that households produce focuses on two components of the waste stream, contributing by weight 57% of the total, namely newspapers and packaging waste. I consider these in turn.

Table 8.3 The household waste stream

Type of waste	Weight (million tonnes)	Percentage of total
Newspapers	2.3	17.7
Packaging waste	5.1	39.2
Paper	1.8	13.8
Glass	1.2	9.2
Plastic	1.1	8.5
Ferrous	0.9	6.9
Aluminium	0.1	0.1
Compostables	3.6	27.7
Other	2.0	15.4
Total	13.0	100.0

Source: Environmental Resources Ltd, 1992.

Newspapers

The UK newspaper industry has, so far, successfully resisted the implementation of value added tax (VAT) on newspapers as a tax on knowledge. The argument is that information or knowledge is a public good, the free dissemination of which is necessary for the effective functioning of a democracy. As a major source of household waste, the issue to be considered is whether there is a case for the implementation of a tax on newsprint to deal with its contribution to the waste stream.

Table 8.4 gives some data on the relative sizes of daily and weekend newspapers. It relates to selected quality broadsheets although the same pattern, but with smaller quantities, is observable with tabloids. The pattern relates to one week in 1995. The pattern revealed is quite striking. The Sunday paper contains almost twice the volume of newsprint of the Saturday paper and in volume terms is almost equivalent to an entire week of daily papers. The cost per unit of newsprint is much lower, less than half, for the weekly papers than for the daily papers. The cost of the Sunday paper, in terms of newsprint, is higher than for the Saturday paper because of other differences in the cost structure: Sunday papers contain a glossy supplement using higher quality and more expensive paper and, having only a weekly outlet, the producer cannot spread the costs of its journalists and print workers in the way that a daily paper can. The Saturday paper is semi-independent and can share some of these costs with its weekday equivalent.

The imposition of VAT on newspapers would raise the price of all newspapers but would not alter the relative prices of weekly and daily papers, either in total, per page of print or per broadsheet of newsprint. This is because VAT is per copy of the paper and not per volume of newsprint. A tax on newsprint, on the other hand, would alter the competitive position of the weekly papers. Table 8.5 illustrates two alternative tax rates of 0.5 and 1.0 pence per broadsheet of newsprint.

While VAT has no effect on the structure of newspaper prices and would only work by discouraging total demand for newspapers, a tax on newsprint has the effect of substantially increasing the prices of the weekend papers in

Table 8.4 The pattern of newsprint consumption

| Newspaper type | Price (pence) | Size (pages) | | Newsprint volume (broadsheets)* | Price per broadsheet (pence) |
		Broad-sheets	Tabloid sheets		
Sunday: *Sunday Times*	100	92	228	103	0.971
Saturday: *Weekend Guardian*	50	60	108	57	0.877
Daily: *Guardian*[†]	45	30	24	21	2.143

* Two broadsheet pages or four tabloid pages equal one broadsheet.

[†] Average over 5 days.

Table 8.5 Implications of alternative taxes on newspaper prices (calculated from Table 8.4)

Paper type	No tax	Percentage of price of daily	VAT @ 17.5%	Percentage of price of daily	Tax of 0.5 p/ broadsheet	Percentage of price of daily	Tax of 1 p broadsheet	Percentage of price of daily
Sunday	100	222	117.5	222	151.5	273	203	308
Saturday	50	111	58.8	111	78.5	141	107	162
Daily	45	100	52.9	100	55.5	100	66	100

comparison with daily papers. The market for newspapers of whatever type is fairly competitive so that it is unlikely that papers would absorb the tax. It would thus provide an incentive for these papers to reduce their size in order to maintain their markets. Environmentally this would therefore be an efficient tax. The regulatory alternative, of imposing a maximum size on papers, would not be an efficient alternative. If it 'bit' with typical sizes, it would prevent papers, if they were to keep within the limits, from increasing their sizes for special events. Furthermore, since there is in fact considerable variation in size for any newspaper type (we have chosen some of the largest), a regulation that constrained the largest papers could allow some of the smaller ones to actually increase in size.

Packaging waste

Packaging waste is the major single component of the household waste stream, constituting almost 40% of the total. While data are not available it seems likely that the growth of packaging goods is a major contributor to the growth of household waste production. Action to reduce it therefore would make a substantial contribution to the problem of waste disposal.

Households are essentially passive recipients of packaging waste. Goods from retail outlets for the most part come pre-packaged and the household discards the packaging. Action therefore is required on the use of packaging materials by manufacturers. Action is rendered difficult since modern retailing is built on pre-packaging; supermarkets are constructed around pre-packaged goods as are a number of other types of retailers including consumer durables, electrical goods, stationery, toys and some parts of clothing. Pre-packaging is an essential part of quality control and commodity standardization. The packaging ensures that the goods have not been tampered with, altered or adulterated at any point in the distribution chain. It gives the consumer the guarantee of quality. With food retailing it is also an important part of ensuring health regulations are met. Attempts at reducing the use of packaging, therefore, could raise retailing costs, making it more labour intensive and result in a reduction in quality of consumer goods.

But the difficulties should not be taken as a counsel of despair. Neoclassical economics would say that if there are negative externalities in goods packaging then they should be internalized to the decision-makers by the imposition of a Pigovian tax. If this is not done then the consumption of packaging will be above the social optimum level and the environment will suffer. We saw in Chapter 4 that this doctrine depends on sufferers perceiving the externality as a nuisance and being able to express their suffering as a willingness to pay. These conditions could not be satisfied for a wide range of current environmental problems with the result that there was no unique optimum state of the environment and in consequence no Pigovian tax. Nonetheless if the environmental problem, in this case that of excessive waste production, is a real one it is open to society to attempt to improve matters by the use of control instruments, one possible instrument being a pollution tax.

Where the consequences of control are to a greater or lesser degree unforeseeable, as in this case, there is a preference for the use of economic instruments, leaving it to the market to determine how the objectives should be met. We have seen that the incidence of economic instruments depends on elasticities of demand. In so far as the costs fall on consumers in the form of higher prices for highly packaged goods, they have an incentive to avoid these costs by modifying their consumption behaviour. Market opportunities are then created for the provision of cheaper goods with less packaging. How this would be achieved, whether through different kinds of retail outlets and an acceptance by consumers of greater quality variation or whether alternative means of ensuring quality are adopted, may be left to the market. The risk of losing out to rivals offering less packaging provides the incentive to manufacturers to examine their practices and to the producers of packaging to look for more economical packaging services.

Despite these arguments for the use of economic instruments the principal instrument currently on offer is the EU Packaging and Packing Waste Directive, which sets targets on producers for the recycling of packing waste. The incentive here is primarily on making packaging easy to recycle but manufacturers can also make their recycling task easier by reducing the quantity of packaging used. However this directive could be seen as technically inefficient in that it is not targeted at waste volume.

The obvious alternative would be a tax on packaging. Packaging is already subject to VAT, which is passed on to the consumer (Box 8.2). Since current VAT levels plainly do not achieve the required level of packaging saving, it could be subjected to a supplement for waste disposal. Unlike the case of newspapers VAT would be an efficient tax since it is targeted at the problem, namely the use of packaging. What the level of the supplement should be is open to question.

Deposit–refund systems

As a final issue in this chapter I consider the application of deposit-refund systems. In many countries deposit-refund systems were traditionally used for beer bottles. The consumer pays a deposit on the container for her drink which she can claim back from the retailer when she returns the empty container. The retailer returns the container to the drinks manufacturer for reuse. If all containers are returned deposit-refund systems do not create packaging

Box 8.2 VAT

VAT is charged as a percentage of value added in a production process. Value added is the difference between the cost of raw materials and other non-factor inputs and the value of the producer's output. Value added therefore is the net revenue of the producer from which sum she pays the wages of the workers, rent on her premises and any charges for borrowing for capital investment. The residue is her profit. Thus value added = wage bill + interest charges + rents + profits. The producer charges VAT as a percentage of her value added and passes it to the VAT authority. The purchaser of the goods, who may be the retailer or, if manufacturing goes through a number of stages, the processor of semi-manufactures, claims back the VAT paid on inputs purchased (in the case of the retailer, on goods purchased) and charges VAT on value added. The final consumer cannot claim back the VAT paid. Thus the incidence of VAT falls entirely on the consumer and on any producer whose turnover is below the limit for VAT registration. Since this limit is very low only very small 'own account' traders are liable to pay VAT. All limited companies are registered for VAT.

waste but inevitably there will be losses, from breakages and failure to return. If the deposit is small then the consumer may not wish to go to the inconvenience of returning. The system then depends on there being people, e.g. children, who are willing to collect the empties for the income that they yield. The deposit cannot be large or it will provide an incentive for stealing drinks for the value of the containers.

The most efficient reuse system is probably the doorstep milk delivery system in the UK. Here the costs of collection are very small since the delivery person collects the empty bottles when she delivers the morning milk. No deposit is required since the demands on the consumer are trivial, no more than putting the bottles on the doorstep.

Deposit-refund systems can only work for suitable reusable containers such as glass bottles and only for products that can be sold in such containers. How far the system can be extended technically, e.g. to perishable foodstuffs sold in plastic containers, is unclear, but for most types of containers deterioration with use is a serious constraint.

Summary

- Domestic waste is a form of non-point pollution with the peculiarity that it is collected and disposed of by designated bodies that are point sources of pollution.
- The principal environmental problems associated with the domestic waste stream are those connected with disposal, especially landfill, and recycling.
- These two problems are interlinked since increasing recycling reduces the volume of waste that has to go to landfill.

- A range of both economic instruments and command and control regulations imposed on the waste-disposal authority can be used to increase recycling and reduce landfill. All of them require that households undertake a preliminary sort of waste and entail that the costs of recycling fall ultimately on the households.
- Recycling costs are positive since the value of many recycled materials is less than the costs of recovery. Nonetheless it may be justified by the externalities of landfill and/or of primary material extraction.
- Markets for recycled materials are limited by the high costs of collection and the quality of the resulting materials.
- None of the available instruments provide any incentive to households to reduce the volume of waste entering the flow since disposal charges levied on households are not related to the volume of waste.
- Incentive-compatible charges are possible but carry the risk that households will resort to illegal dumping of waste.
- The principal components of the household waste stream are newspapers, packing waste and compostables.
- Households with gardens might be persuaded to undertake some of the composting. Reducing the flow of newspapers and packing waste is most efficiently dealt with by instruments directed at the producers. Economic instruments such as taxes on newsprint and on packaging are probably more efficient than regulations.

Chapter 9

Atmospheric emissions from road vehicles

The problem

Road transport is a major source of air pollution at the local level and a substantial contributor to global atmospheric pollution. The principal polluting emissions are as follows (data derived from Department of the Environment, 1992):

Carbon monoxide (CO): 90% of all UK emissions
Nitrogen oxides (NO_x): 51% of all UK emissions
Volatile organic compounds (VOCs): 41% of all UK emissions
Particulates (black smoke): 46% of all UK emissions of which about 40% is
    attributable to diesel vehicles
Carbon dioxide (CO_2): about 20% of all UK emissions
Lead: additives in petrol are the major source of airborne lead.

In addition road transport is a minor but significant contributor to other airborne pollutants such as sulphur dioxide (2% of UK emissions) and nitrous oxide (7% of UK emissions). As a major contributor of the ingredients (VOCs and nitrogen oxides) road transport is an important source of tropospheric ozone concentrations.

It is clear from these data that any programme to control atmospheric pollution has to address the problem of pollution from road transport. In fact these data substantially understate the role of road transport for two reasons:

1. In several cases road transport is effectively the sole source of *growth* in atmospheric emissions. Thus road transport is the source of a 30% growth in CO over the last decade and was the main cause of a 35% increase in NO_x between 1986 and 1991. It is also the sole source of growth in VOCs. With black smoke and CO_2, road transport emissions have been growing while other sources are declining. CO_2 emissions from road transport doubled between 1970 and 1990 and are predicted to show further substantial increases in the next 25 years, being the strongest growth sector over that period. By 2020 CO_2 emissions from road transport will exceed that from electricity generation and other industry to become the largest source of emissions.
2. Road transport plays a much larger role in local air pollution of major urban areas. As an example the contribution of diesel engines to emissions of black smoke is 80% in urban areas.

Vehicle emissions are a major impediment to the UK meeting its obligations under various EU directives on air quality and atmospheric emissions and to reducing the emission of greenhouse gases under the Climate Change Convention. Many people in the UK, particularly in urban areas, are subject from time to time to exposure to concentrations of air pollutants that exceed World Health Organization (WHO) guidelines. This has been the case with the secondary pollutants of nitrogen dioxide and tropospheric ozone and with one size of particulates known as PM10. Vehicle emissions are usually a major factor in these incidents. Air quality is a factor in the incidence of respiratory disorders and is suspected of being the root cause of the sharp increase in incidence of asthma among children. With the current state of medical knowledge there is uncertainty as to which substances or combination of substances are responsible for this phenomenon. The exact role played by vehicle emissions thus cannot be determined. Benzene, a component of vehicle emissions, is a suspected carcinogen. Diesel fumes are particularly suspected of being carcinogenic.

Of course air pollution is not the only environmental problem posed by road transport. It is the largest single cause of noise pollution, disrupts local communities and is a cause of deaths through traffic accidents. The construction of roads adds another series of environmental concerns, which are briefly considered in Chapter 12.

The causes

The major reason for the high and growing contribution of road vehicles to air pollution is the increase in the volume of traffic. This has grown by over 600% since the early 1950s and in 1990 road transport accounted for 93% of passenger transport and 81% of freight moved. Overwhelmingly the largest component of traffic growth has been that of private cars, which in terms of vehicle kilometres has increased in the 40 years from 1950 by over 1000% to 330 billion vehicle km. Road freight traffic growth has, by comparison, been relatively moderate with ton kilometres of goods carried increasing by only 140% over the same period. For this reason, and because the subject of road traffic and air pollution is one of great complexity to which this chapter can be no more than a brief introduction, I concentrate on the private motor car.

Typical journey lengths in the UK are very short; in 1989–91 over one-third were less than 1 mile and three-quarters were less than 5 miles (the official statistics define a journey as any course of travel of greater than 200m not undertaken on foot). Journeys of under 25 miles account for over half of the average distance travelled per person each year and for 96% of the total journeys undertaken. The motor car is the dominant mode of transport and accounts for 74% of all journeys of greater than 1 mile within the UK; 50% of urban journeys within the London area and up to 65% of urban journeys in other UK cities are by private car.

The use of the car for short journeys is a critical factor in the contribution of road transport to problems of air pollution for four reasons:

1. Per person mile, private motor cars are far more polluting compared with public transport modes and, of course, alternative private modes of walking and cycling.
2. They are concentrated in urban areas where the local effects of pollutants are greatest.
3. The slow speeds and frequent stops and starts means that engines operate inefficiently, increasing emissions of all pollutants including global CO_2.
4. Exhaust gas temperatures do not reach levels at which pollution-control devices (e.g. three-way catalytic converters) work efficiently, if at all.

There are three broad lines of approach to deal with this problem:

1. Accept consumer choice about journey frequency and modal split and reduce the volume of air pollution that results from this level and pattern of car use.
2. Accept the volume of journeys but seek to alter the modal split towards less polluting modes, i.e. public transport, cycles and walking.
3. Reduce the volume of short-distance journeys undertaken.

These are of course not mutually exclusive alternatives and any strategy for dealing with air pollution from road transport will involve a combination of all three approaches. However it provides a useful framework for considering policy options.

Air pollution and vehicle technology

Emissions limits for cars are specified by EU standards set for carbon monoxide, hydrocarbons and nitrogen oxides and particulates. These standards are applied in three stages. Stage I was introduced in 1993. Stage II will come in 1997 and will lower emissions limits. Stage III will operate from 2000 and will tighten limits still further. There is as yet no agreement as to what the Stage III standards will be. Emissions standards are specified separately for petrol and diesel engines. In addition to these limits controls on CO_2 emissions will probably be needed to meet national responsibilities under the Climate Change Convention.

The main technological options for tackling the problem of air pollution can be listed as: cleaning exhaust gases; improving fuel efficiency; and changing the means of propulsion.

Cleaning exhaust gases

The principal means of cleaning exhaust gases is by the fitting of catalytic converters. Closed loop three-way catalytic converters are currently the only way of meeting Stage I standards and have been required to be fitted to all new cars since 1993. Catalytic converters face four problems:

1. There are no requirements for retro-fitting of catalytic converters to existing cars and for this reason the pace of air-pollution improvements is

limited. The existing fleet of cars is replaced over a period of 12–15 years and it will take at least 5 years before 50% of the fleet is fitted with catalytic converters. It has been estimated that, by 2000, 30% of cars will have been manufactured to Stage II standards, 40% to Stage I standards and 30% will still not have a catalytic converter. A policy option to speed up the process would be to tighten emissions standards required at the annual testing of vehicles of over 3 years old. The objection to this is that it would be regressive in its impact since older and cheaper cars are acquired by the poorer sections of the community.

2. They only operate when exhaust gases reach a certain temperature and a high proportion of trips, almost 50% according to one study, are too short for catalytic converters to operate efficiently.
3. They reduce engine efficiency and in consequence increase emissions of CO_2.
4. They do not work for diesel engines.

Improving fuel efficiency

CO_2 emissions are closely related to fuel consumption. Fuel economy is influenced by a number of factors including the standard of maintenance of vehicles and the way they are driven. Petrol engines are at their most efficient at speeds of 60–80 km/hour and efficiency declines rapidly at higher and lower speeds. Efficiency also falls with acceleration and is particularly low when idling. Diesel engines generally have lower fuel consumption and reach peak efficiency at speeds of 50–60 km/hour. Both petrol and diesel engines (the latter at least as fitted to cars) are therefore highly inefficient in congested urban conditions. Fuel efficiency requires that vehicles go faster in urban areas and more slowly on motorways. Alternatively separate vehicles are required for the two conditions, each designed for its function. Electric vehicles rather than internal combustion engines constitute appropriate technology for short-distance urban journeys. However, it is not realistic to expect that people will be willing to own and operate two vehicles. Apart from problems of parking and the loss of flexibility entailed in having to fetch your other vehicle when you want to use it, the cost would be prohibitive.

The other major factor affecting fuel efficiency is the size of vehicles and their power–weight ratios. Weights of specific models of cars have risen over the last 30 years because the impact of the incorporation of additional design features for comfort and safety has more than offset trends towards the use of lighter materials (aluminium and plastics). Vehicle size is subject to swings of fashion. Fuel economy rose in the 1970s with the growth of minis and small hatchbacks but has recently declined with trends towards larger and heavier four-wheel drive vehicles. The obvious instrument for reversing this trend would be differentiation of excise duty in favour of smaller engine capacity. It has been estimated that up to 50% reduction in CO_2 emissions could be obtained if drivers were prepared to accept smaller and lower performance cars.

Changing the means of proplusion

Gains to fuel economy can be obtained by switching from petrol to diesel engines but the resultant savings in CO_2 emissions are at the expense of increases in particulates and other pollutants. Technical opinion has swung away from the diesel, which is now considered undesirable on health grounds in urban areas. Electric cars are certainly less polluting and are efficient in urban driving conditions, but the technology for the large-scale replacement of the internal combustion engine with electric motors is not at present in place.

While technological improvements are desirable *per se*, their role in dealing with the problem of air pollution is limited by two factors: the long lead-in time, because the technology is incorporated only in new vehicles; and the problem of traffic growth. The Department of Transport produces two forecasts of future traffic growth: a high and a low growth forecast. These are respectively 142% and 83% between 1988 and 2025. While the forecasting methods used are a matter of dispute, it is clear that any improvements in air pollution due to vehicle technology are swamped by the sheer growth of traffic. Technology is having to run very fast just to stand still. Furthermore the technology does nothing to address the numerous other environmental problems of traffic growth. Hence the other approaches of changing modal split, and reducing total volume of journeys undertaken by any form of vehicle other than the bicycle or the horse and cart are necessary if the problem is to be seriously addressed.

Choice of transport mode

The growth of car transport for journeys has, of course, been at the expense of other transport modes: cycling and the public transport alternatives of bus and rail, all of which have declined over the post-war period. As a transport mode the private car can be described as a superior product, offering a combination of comfort, flexibility, convenience and (apparent) security that cannot be matched by the alternatives. These aspects of product quality determine the decision of households to own cars. Car purchase is a large element in the average household budget and is often spread by credit arrangements. But however it is paid for, once the ownership decision is made, these fixed costs are 'sunk' and should not affect the choice of transport mode for a particular journey, which for the rational consumer should be decided by a comparison of the variable costs of the alternatives (Box 9.1). The cost of car purchase is not the only fixed cost of motoring; vehicle excise licence and car insurance must be incurred before the car is put on the road and thereafter are independent of the use made of the vehicle. If the car is used for any journeys these costs are also sunk costs. The variable costs of use are merely fuel costs and the element of depreciation that is use-related, i.e. wear and tear. Thus, as a transport mode, the private car is characterized by high fixed costs and relatively low variable costs.

This cost structure encourages car owners to maximize the use they make of their vehicles and biases modal choice for car owners towards car use since public transport fares have to be set to cover the fixed costs of the operation including capital expenditure on vehicles, staff wages, vehicle licence fees, etc. Hence the motorist is faced with a comparison between the marginal costs of a car journey and a public transport fare based on average cost. In the case of rail the cost of track provision is also averaged in the fare. With road transport it has been a long-standing practice for the levels of taxation, vehicle excise duty and fuel tax, to be set to cover annual expenditure on road construction and maintenance. Thus the cost of track provision for bus transport is also averaged in the fare. Since total tax receipts are set to cover the costs of road provision and approximately 20% of taxes paid by the average private motorist is in the form of the fixed cost of excise duty, it could be argued that further bias in mode choice is imparted by the fact that only 80% of track costs appears as a marginal cost to the motorist.

This situation is aggravated by the problem of company cars of which there are approximately 3 million in the UK, about one-eighth of the total. Drivers are taxed on the imputed benefit from company cars but there seems little doubt that the full benefit is not recovered and that the tax provisions themselves tend to encourage maximization of use and militate against choice of

Box 9.1 Concepts of cost

Total cost of production can be classified into two components: fixed costs, which do not vary with output, and are incurred if any output at all is produced; and variable costs which vary with output. Economics usually works with two cost concepts derived from these: average costs, defined as total costs divided by output, and marginal costs defined as the addition to total costs that results from a unit increase in output. Provided that the addition to revenue received by the producer from an increase in output exceeds the marginal costs of producing it, then the increase in output adds to total profit and makes a contribution to fixed costs. Thus a rational profit-maximizing producer will expand output as long as the receipts exceed marginal costs. The importance of average costs is that, for production to be profitable, total revenue must exceed total costs. Since the total revenue equals the price received per unit times the number of units sold, price must exceed average cost in profitable enterprises. Thus a bus company must have a fare structure that ensures that the average fare paid is at least equal to its average cost. However if it can capture additional passengers by offering special concessionary fares, which at least cover the marginal costs of carrying those additional passengers, it is in its interest to do so. Once the decision is taken to produce any output (e.g. to operate a bus service) then the fixed costs are incurred and cannot be recovered. These costs are then said to be sunk. They then should not affect decisions about the level of service.

alternative modes. About 20% of employees with company cars also receive free petrol. While again the imputed benefit is taxed, the taxation is reduced if the driver travels more than a threshold value of business mileage. This is thought to encourage unnecessary business trips in order to avoid tax.

A number of instruments have been suggested to encourage the use of alternative transport modes by motorists. Economic instruments include the following:

1. raising the level of fuel tax;
2. restructuring the taxation system so as to make all payments vary with distance travelled;
3. road pricing;
4. subsidies to public transport.

In addition there are a number of proposals for physical controls to make short-distance urban journeys less easy for the motorist. I consider these in turn.

Fuel tax

Raising the taxation on petrol and diesel is the most straightforward and obvious instrument for discouraging motoring. If motorists are rational it should have the greatest impact on short-distance journeys where fuel efficiency is at its lowest and where alternative modes are the most easily available. The UK Government is committed to raising the excise duty on petrol by at least 5% a year in real terms as a central plank of its strategy for reducing CO_2. The Royal Commission on Environmental Pollution (1994) suggested a 9% per annum increase as part of its strategy for meeting targets for reducing air pollution from road transport.

The effectiveness of fuel tax as an instrument depends on the elasticity of demand for petrol and diesel. Department of Transport research suggests that the elasticity in the short term is fairly low, a 10% rise in the price of fuel resulting in a 3% fall in fuel demand of which about half would result from reduced vehicle use. On this basis the price of fuel would have to rise in real terms (i.e. above the rate of inflation) by 80% by 2000 and 250% by 2025 to meet the Government's target of limiting CO_2 emissions from road transport to 1990 levels.

It is likely, however, that the elasticity would be higher over the longer term as drivers have the flexibility to change their patterns of travel and choose more efficient vehicles. One estimate is that in the longer term a 10% rise in fuel prices would result in a 7% fall in fuel demand (Goodwin, 1992).

'Pay at the pump'

I have argued above that modal choice is biased against public transport by the element of fixed costs of motoring. The suggestion here is that these fixed costs should be converted into variable costs by recovering them through the tax on fuel. This could obviously be done with the element of fixed cost that is taxation, the annual vehicle excise duty. This would add approximately 20% to fuel tax. The argument against this is that the purchase of the annual licence is used to ensure that the legal requirements for third party insurance and for testing of vehicles are complied with. The annual vehicle test incorporates a test of exhaust emissions and a loss of the capacity to enforce this could adversely affect the problem of air pollution. However the excise licence could certainly be reduced in cost to a nominal value and the burden shifted to fuel tax without losing these benefits.

Road pricing

Fuel tax only discriminates between types of journey in so far as motorists recognize that fuel efficiency is lower and hence journey cost is higher in congested urban conditions. At best it is a crude discriminator. The idea with road pricing is that much finer discrimination is achieved and that the price of journeys that are particularly undesirable on environmental grounds, such as commuting in urban areas, and ones where it is thought that modal change is possible, is raised to the level necessary to reduce the traffic. This concept of road pricing is different in objective from the charging of tolls on motorways and bridges, which are intended to recover the costs of construction and, with current UK policy, to attract private capital into building roads.

Road pricing to restrict traffic in cities and urban areas is dependent on charging technology. Toll booths are thought not to be feasible because of the congestion that they would cause at peak times, when it is desired to price traffic out of the city. Instead the idea is to fit vehicles with devices that allow for automatic charging. Over the last decade the message in all studies that have considered road pricing has been that the technology exists and that large-scale trials to test it will shortly take place. Road pricing thus remains imminent technology.

Currently practical alternatives include parking charges and pre-purchased entry tickets to controlled areas, which have been used for some time in Singapore. The major drawback of all road pricing schemes is that they stand in danger of diverting traffic and congestion to areas not subject to control. They can, however, work for traffic such as commuting whose destination is in the controlled area. Park-and-ride schemes for commuters and shoppers, where cars are left in car parks at city edges and the parking ticket includes the cost of a dedicated bus to the city centre, are used in a number of towns and cities in the UK.

Subsidies to public transport

An alternative to raising the price of car travel is to lower the cost of public transport through subsidies. Subsidies have fallen out of favour in the last 20 years. The objections to them are, first, that the incentive to efficiency in public transport is removed so that the subsidies can be absorbed by increases in cost rather than reductions in fares and, second, that they are wasteful in that fares are lowered for those who would use public transport in any event and not merely for those who switch from cars. Nonetheless subsidies for urban rail transport and underground systems are widespread and probably necessary for their viability.

Physical controls

The problem of traffic in towns, with its attendant air pollution and other environmental problems, can probably only be controlled by physical restrictions used in conjunction with other measures. The physical controls include restrictions on traffic access by pedestrianization and traffic loop systems, restrictions on the availability of parking, bus priority lanes and junctions. One objective of such schemes is to make the use of public transport faster and more convenient than the use of the car, thus offsetting for short-distance urban journeys the car's advantage of convenience and flexibility.

Reducing the volume of journeys

Many of the instruments discussed in the preceding section will have some impact on the volume of journeys undertaken. What remains to be briefly discussed in this section is the impact of urban form and settlement patterns on the volume of journeys.

In the long run the urban structure adjusts to the mode of transport and the pattern of mobility that it facilitates. The long love affair with the motor car has led to modification of urban structure to exploit the advantages of the motor car. Examples are the development of out-of-town shopping complexes dependent on the car as a means of transport and the increasing separation of rural or semi-rural residence and urban employment. There is evidence that concentrated urban structures reduce the demand for short-distance movement. Urban trends are in the other direction and the inner city is losing its previous economic function with the high street declining in importance as the retailing centre and increasingly employment moving out as well.

The burden of control for this aspect of the problem falls on the planning system and solutions are necessarily long-term ones. Whether in fact it is possible to reverse previous causality and to modify the urban form towards reduced needs for mobility is a very large question that I will not even attempt to answer.

Summary

- Road transport is a major source of both local and global air pollution and gives rise additionally to a host of other environmental problems.
- Private motoring is the principal cause of these problems and the growth of private motoring represents one of the major challenges to sustainable development.
- Typical private car journeys are short and concentrated in urban areas.

- Feasible gains in air quality from the technical improvement in vehicle performance are swamped by the sheer growth of urban traffic. Hence practical solutions entail the reduction of car use in urban areas.
- Private motoring is subject to relatively high fixed costs but low variable costs. Hence it is rational for car owners to maximize use of their vehicles.
- Economic instruments work by shifting the balance between fixed and variable costs of motoring. Possible instruments include increasing fuel taxes and collecting vehicle licence duty and insurance 'at the pump'.
- Road pricing in urban areas is an alternative approach. It is possibly technically feasible but there remain many problems. Road pricing always carries the risk of avoidance by motorists using unpriced alternative routes and thus transferring the location of the problem rather than treating it.
- Physical control via restrictions on vehicle use in urban areas, and particularly on commuting by car, are a necessary and probably a major part of policy to deal with the problem in the short term.
- The problem is made intractable by the fact that the urban structure has adjusted to accommodate intensive mass private car use. In the long run, sustainable solutions involve redesign of the city. But this can only be a drawn-out and expensive process.

Part 3

Cost–benefit analysis

Chapter 10

Principles of cost–benefit analysis

Domain of cost–benefit analysis

The discussion in the last few chapters has been concerned with the impact of current production on the environment. In this chapter I turn to the issue of investment decisions. Cost–benefit analysis (CBA) is used for the appraisal of public sector investment projects and other aspects of public policy. It is used, for example, for the appraisal of infrastructure projects such as the building of roads, coastal defences and power stations. But it can also be used to assess policy initiatives such as the implementation of policies for the control of specific types of pollution or to encourage the recycling of waste.

Under CBA the total social benefits anticipated from a project are compared with the social costs and a decision is taken on the project by the use of the decision rule: invest if the present value of benefits exceeds the costs. Environmental effects are included among the costs and benefits.

Whose costs and benefits?

In the conduct of a CBA no account is taken of the incidence of the costs and benefits. The costs of a project may be borne by one set of households and the benefits accrue to an entirely different set. This will not matter; the thinking is that the project is undertaken by society for society's benefit and all costs and benefits are treated equally provided that they are internal to society. Under restrictive assumptions, satisfying the decision rule for CBA will result in a potential Pareto improvement, i.e. the gainers from the project could compensate the losers and still be better off. But there will not be an actual Pareto improvement unless the gainers actually do compensate the losers. CBA practice does not require that compensation is paid.

The incidence of costs and benefits from public sector projects is often highly uneven. Thus it has been shown that road building projects benefit middle and upper income groups at the expense of the poor (Pearce and Nash, 1981). The recognition of the distributional problem of CBA, i.e. the incidence of costs and benefits is uneven between different income groups, has led some economists to suggest that CBA rules should be modified to take account of distributional factors by the incorporation of distributional weights into measured costs and benefits. (Pearce and Nash, 1981, pp. 33–37). This procedure would greatly complicate the conduct of CBA and there is furthermore no agreed set of weights that should be used for the purpose. In any case it can be argued that distributional weighting of CBA is not necessary for two reasons:

- Because of the operation of swings and roundabouts. There are a great many public sector projects each, or certainly each type, with a different distributional impact. Every group in society gains from some projects and loses from others. The distributional impact of the complete set of projects is much fairer between different groups in society than might be supposed from examination of individual projects in isolation.
- The Government is the arbiter of distributional issues and, through its taxation and expenditure policies, has the power to alter the distribution of income between its citizens. If therefore the Government deems that the distributional effects of a project are unacceptable it can correct the problem through its fiscal policy. If it chooses not to do so then this is because it judges the distributional consequences not to be serious. The most that the CBA analysis needs to do is to provide information about the distributional implications.

Since distributional weightings are not used in CBA practice I do not discuss the matter further.

Problems arise where some of the costs and benefits are external to the society, where for instance a project impacts on trans-boundary or global pollution. Normal practice has been to ignore external costs and benefits, which can lead to some bizarre decisions. An example involving CBA of land drainage is discussed in Chapter 12. With increasing international concern about the global environmental impact of economic policies, ignoring external impacts in project appraisal might be thought to be unacceptable. External effects will however be taken into account if the country is committed by international treaty or other agreement to protect the international environment. Countries within the European Union (EU) have a number of obligations and membership of the Organization for Economic Co-operation and Development (OECD) provides others. This issue is discussed further later in the chapter and in Chapter 15.

A related problem of excluded benefits concerns CBA and benefits accruing to future generations. Discussion of this issue is postponed until we come to discuss sustainable development.

Investment appraisal

CBA is a technique of investment appraisal for the public sector. It builds on an analogy with what might be called best-practice investment appraisal for the private sector and justifies departures from that best practice on the grounds that the investor is society as a whole. The idea can be conveyed by starting with a simple example of private investment decision-making.

A factory owner is considering whether to buy a machine that will cost £1000. If she buys the machine she can increase her output over the next 5 years, after which the machine will be worn out/obsolete and will be thrown away. She estimates that she can sell the additional output, net of the additional costs of labour and materials etc. that she will incur, and after any taxes that she must pay, for £500 in the first year and for £200 in each of the following 4 years. This is her profit from the investment in the machine. There are two alternative situations that she might be in:

1. She has the required capital. In this case the alternative use of her £1000 would be to put it in a bank to earn interest. If she leaves it to accumulate then, by compound interest, at an interest rate of $r\%$, after 1 year she will have £1000 $(1 + r)$, after 2 years £1000 $(1 + r)^2$ and after 5 years £1000 $(1 + r)^5$.
2. She has to borrow the capital. In this case she will have to find interest payments of £1000r for each of 5 years and has to pay back the capital sum of £1000 at the end of the period.

Provided that the interest rate on borrowing was the same as the rate paid on lending it would be a matter of indifference whether our entrepreneur borrowed or used her own capital. This is clear if we consider the situation where she places £1000 in the bank and borrows another £1000 to buy the machine. The money in the bank would yield just enough interest to pay the interest on the loan and at the end of the period she could close her account and pay off the loan.

She will not be indifferent between the two options because she will have to pay a higher rate of interest on borrowed money than she would receive on money lent, the difference between the two representing the costs and profits of the financial institutions accepting deposits and making loans. Because of this, the interest rate that she will use for appraising the investment will vary with the means of finance. If she has to borrow the capital sum she will appraise the investment at the higher borrowing rate. If she does not, the lower lender's rate is appropriate. Having raised this minor complication we will now ignore it and assume that borrowing and lending rates are the same.

To determine whether the machine purchase is her best strategy our potential investor calculates the *present value* (PV) of the extra profits from the investment as follows:

$$PV = 500/(1 + r) + 200/(1 + r)^2 + 200/(1 + r)^3 + 200/(1 + r)^4 + 200/(1 + r)^5$$

where r is expressed as a decimal (thus 1% is expressed as 0.01).

If this sum is greater than £1000 then she will get a better rate of return on her money by buying the machine than by putting the money in the bank. This is the process of discounting. $(1 + r) \ldots (1 + r)^5$ is the *opportunity cost* of the investment; it is the interest forgone by buying the machine. If the profits, when weighted by the interest forgone in earning them (or the interest that has to be paid in order to earn them if the capital sum is borrowed), exceed in value the capital committed, then the investment is worthwhile; it yields a higher rate of return than the alternative use of the capital sum.

Thus our industrialist uses the PV decision rule to appraise her investment. That rule is as follows: discount the anticipated profits at the opportunity cost of the capital (the interest rate) and invest if the PV of the discounted profit stream exceeds the capital cost.

Our industrialist could instead have used an alternative decision rule for appraising her investment. Having estimated her anticipated profit stream she could have found the discount rate that would make the PV of that profit stream equal to the capital cost. Thus she would look for the discount rate i that would solve the following equation:

$$1000 = 500/(1 + i) + 200/(1 + i)^2 + 200/(1 + i)^3 + 200/(1 + i)^4 + 200/(1 + i)^5$$

Invest if $i > r$. In this equation i is known as the internal rate of return (IRR) on the investment and the decision rule is the internal rate of return rule.

With our hypothetical investment the two rules will obviously give the same answer. If the profit stream when discounted at r has a PV greater than the capital cost (£1000) then the discount rate that will make PV = £1000 must be greater than r. IRR gives the industrialist the actual rate of return on her investment; the PV simply tells her that that return is greater than the opportunity cost of her capital committed. The PV is the easier to calculate but it yields less information.

But the two decision rules will not always give the same result. Figure 10.1 gives the net cash flows (receipts minus expenditure) from the project. Initially (year 0) the cash flow is negative (–£1000). Thereafter for the next 5 years), constituting the life of the investment, it is positive. Now suppose that there is a year 6 in which the cash flow becomes negative again, perhaps because she has to dismantle her machine and decontaminate the site. This negative flow is shown as a dashed line in the diagram.

Obviously the year 6 commitment will reduce the PV of the project but it will have the effect of creating two values for the IRR. This is because the discount factors, $1/(1 + i)^n$, constitute a set of declining geometric weights. As the discount rate is raised so the weight given to cash flows, profits or expenditures falls relative to cash flows that are closer in time. Thus at a discount rate of 1% a cash flow in year 6 is weighted at 94% of its expected value. This weighting falls to 56% with a discount rate of 10% and to 33% with a discount rate of

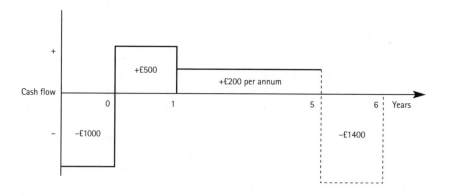

Fig. 10.1 Cash flow from a hypothetical investment.

20%. At high discount rates the initial expenditure dominates the calculation and future cash flows are irrelevant. At intermediate rates the expected profits in years 1–5 are dominant; at very low rates the cost of clear-up becomes significant and the project is again unprofitable. Thus the example given will have two IRRs: one low and one high. The equation for the IRR is a polynomial and will have a real solution (i.e. a positive value of i) every time the cash flow changes sign. Since our example changes sign twice, between years 0 and 1 and between years 5 and 6, it will yield two IRRs. If the opportunity cost of the investment, r, happens to lie between these two values ($i^1 < r < i^2$) the IRR decision rule is giving ambiguous answers. The PV rule however is always unambiguous since it uses only one discount rate, the opportunity cost of the capital expenditure. A real example of this sort of problem would be a nuclear power station that has a high initial capital cost and large decommissioning costs once the plant has reached the end of its life. IRR would not be a useful decision rule for this type of investment.

CBA is an equivalent investment appraisal to the above example for a public investment. It differs from the appraisal of our hypothetical industrialist in the following ways:

- the treatment of transfer payments;
- other forms of price distortions;
- the treatment of externalities;
- the treatment of risk;
- the choice of discount rate;
- the treatment of public goods.

These are discussed in turn.

Treatment of transfer payments

In estimating the revenues received from the investment the industrialist will deduct any tax to be paid. In calculating the cost of the investment she will equally deduct any subsidies she receives. The investment appraisal is made net of taxes and subsidies. From the viewpoint of society the tax paid is part of the benefits: it is a transfer of purchasing power from the industrialist via the Treasury to others. Equally, any subsidy paid is part of the costs of the project: it is a transfer of purchasing power from society to the industrialist conditional on it being spent on the investment. *Thus society would measure the benefits gross of, not net of, the distributional taxes and subsidies* (income taxes, corporation taxes, investment grants and allowances). This difference in treatment can be important in heavily subsidized industries such as agriculture and forestry. Thus in projects that increase agricultural output, such as rural land drainage and flood protection schemes, the value of the additional output to the farmer can be two or three times the social value and it is not uncommon for all additional profit to the farmer to be made up of consumer and taxpayer subsidies. CBA of agricultural land drainage and flood protection is discussed in Chapter 12.

Taxes levied to correct externalities such as pollution taxes are not distributional taxes and should not be grossed out in CBA for reasons given below.

Shadow prices

The opportunity cost of any resources used by a private investor is the price that has to be paid for them. Properly functioning competitive markets ensure that these resources are available at the lowest opportunity cost. If markets do not function properly then the price paid for resources, factors of production and goods and services, may not correspond to the opportunity cost to society of providing them. If this is the case then *in CBA the social opportunity cost should be used and not the market price.* Since this social opportunity cost is not a price that is actually observed in a market it is termed a *shadow price.* In the presence of market distortions CBA should be conducted with shadow prices.

The most pervasive source of price distortions is monopoly. The monopolist extracts a monopoly rent from the consumer by charging more than the competitive price for what is provided. The competitive price is the social opportunity cost of goods and services provided. The competitive price cannot of course be observed in monopolized markets and has to be estimated. However this is a difficult and contentious task in a modern economy since many markets are subject to the influence of monopoly and it may well be that the inputs used by the monopoly producer of a product are themselves distorted by monopoly pricing. Indeed to calculate shadow prices it would be necessary to estimate what the complete set of demand and supply curves would be were all markets competitive! This is not a practical proposition and the problem of monopoly is therefore ignored in CBA, which is conducted in terms of market prices. There are, however, a few areas where shadow prices are deemed to be practical and are sometimes used.

Unemployment

One of these concerns the valuation of labour in the presence of unemployment. If, through its investment, a private company provides work for individuals who would otherwise have been unemployed, that employment is a cost to the company and its opportunity cost is the wages paid. Society's view on it, however, may be different. For a start it is not interested in the current labour force status of the individual workers concerned but only on the impact of the investment on aggregate unemployment. Thus, if giving work to the unemployed results in others currently in employment becoming unemployed, then the social opportunity cost of the workers employed on the project is the same as the private opportunity cost. However if the project reduces unemployment in total then the social opportunity cost of the new employment is not equal to the wage rate since in the absence of the project some workers would have been producing nothing. Society will save on unemployment benefit and/or income support, which is of course simply a transfer payment and is therefore ignored, but it will gain from the output of the newly employed workers. If the project adds to total employment in the economy, by reducing total unemployment, then the social opportunity cost of the workers is zero[1].

Whether a given investment project makes a net reduction to unemployment is not immediately apparent from inspection of the project. This is so even with projects whose principal purpose is to create jobs. Thus a recent public sector project in the UK has been the construction of a tidal barrage across the mouth of Cardiff Bay. This project caused considerable controversy since it involved the loss of tidal feeding grounds for waders in an estuary of international importance. The objective of the scheme was the regeneration of the depressed inner urban area of Cardiff around the bay. The barrage turns Cardiff Bay from a tidal estuary into a large lake. According to the proposers, the Cardiff Bay Development Corporation, the barrage will provide a major positive environmental feature with the removal of unsightly mudflats and their replacement with the pleasing prospect of a large expanse of permanent water. This in turn will attract industry to the area leading to its regeneration and creating a substantial volume of employment. But it has to be asked where this employment will come from. It is clear from a substantial number of studies of inner city regeneration that much of the new industry is in fact short distance moves of existing firms to the new sites. In the case of Cardiff Bay this would be moves from adjacent Welsh valleys, themselves depressed, and from other parts of Wales. That is, many of the jobs would not be employment creation but relocation of existing employment. Even where the jobs were created in wholly new firms that had not previously existed elsewhere, the question has to be asked what would have happened were the Cardiff Bay Barrage not built. Some of these new firms would then have been set up in other locations, perhaps in the south of England, or in the North or in Scotland or Ireland. In fact apart from temporary jobs in the construction industry in building the barrage and new premises for relocated firms, the presumption is that the scheme will create few if any new jobs, since all that it does is to increase the attractions of Cardiff Bay relative to other locations[2]. Thus the fact that it will reduce unemployment in Cardiff is no reason for placing a lower shadow price on the jobs.

While the distinction between local and national reductions in unemployment is clear with projects for urban and regional development, the issue is less obvious for other types of project such as power stations and trunk roads where the intention is not principally to benefit a local area. Even with national projects, of course, the employment benefits to local communities are usually cited as an argument for the project. CBA does not recognize these as

---

[1] Except in so far as the unemployed derive some benefit from the leisure that unemployment yields them. The presumption must be that the unemployed would prefer less leisure and more income; this is implicit in the requirement for the receipt of unemployment payments that they be seeking work. The social opportunity cost of additional unemployment would only be greater than zero if the project resulted in the newly employed working more hours than they would wish to do.

[2] Since CBA takes no account of extra-territorial impacts, the scheme will create new jobs if it attracts employment that would otherwise have gone abroad but this will be a very small part of the total employment created. Even with new foreign firms coming to the area of the barrage it would have to be shown that in the absence of the barrage these firms would not have gone to some other part of the UK. Since all local government in the UK is competing to attract foreign investment to its area this is hard to prove.

benefits unless they involve the creation of additional employment. In all other cases the employment is treated as it would be by industry, i.e. as a cost.

The issue of the employment effects of public investment is one of macro-economics not the microeconomics that I have used in this book. Current doctrine holds that public investment does not in general create employment since it tends to displace private investment of equal value. In UK practice, therefore, employment is not shadow priced in CBA. The exception is where the unemployment is said to be structural. An example where shadow pricing of unemployment was practised was the 1972 Treasury study of forestry policy. It was argued that public sector forestry created employment in remote rural areas and the people affected either could not afford to move to the towns where there might have been work for them, or that as a matter of policy it was undesirable that they should do so since it was wished to maintain rural communities. Forestry projects in this case would add to total employment. This assumption was subsequently challenged (for a detailed critique see Bowers, 1979) and the exercise has not been repeated. The Treasury itself considered that the cost per job created by forestry compared unfavourably with alternative job creation strategies.

Exchange rates

The major area of CBA where shadow pricing is practised concerns exchange rates for the currencies of Third World countries. In a properly functioning exchange market a country's exchange rate (or rather set of exchange rates; it is simpler, and does not destroy the essence of the argument, to write as though there was only one exchange rate) depends on the supply and demand for its currency. The supply will come from its citizens wishing to buy foreign currency in order to import foreign goods, to travel abroad and for other reasons. The demand will come from foreigners wishing to acquire its currency in order to buy its goods, to travel in the country and so on. Countries often interfere in the exchange market in order to protect the exchange rate of their currencies and to influence their balance of payments. They can do this in a wide variety of ways: by exchange controls, placing restrictions on the ability of their citizens to buy and sell currency but also through trade policies, inhibiting imports and encouraging exports.

Where a public investment project has implications for international trade it may be necessary to appraise the project at a shadow exchange rate representing the real value of exports and imports to the society concerned. Some commentators go further and argue that in CBA a shadow exchange rate should be used in any event to give a higher value to exports and a lower value to domestic consumption than would be given by the country's domestic price structure. This stems from a theory of economic development that sees exporting activity as the key sector for development.

Shadow pricing of foreign exchange is not an issue for environmental economics and I do not discuss it further.

Treatment of externalities

If the investment produces externalities the industrialist will ignore them unless they are internalized to her decision-making; for example, if an investment causes pollution that would be internalized to the decision-making process by the levying of a pollution tax or the operation of a command and control system. But externalities to the individual are internal to society and must be taken account of in CBA.

In the context of investment decisions there are two sorts of externalities to be considered: *flow externalities*, which are continuing consequences of the decision; and *impact externalities*, which are one-off consequences of the decision. Flow externalities are analysed in Part 2 (Chapters 4–9). Impact externalities are factors such as the destruction or damage to wildlife sites or to landscape from the investment decision. Examples might be the building of a road through a nature reserve or the intrusion of a power station on a rural landscape. CBA attempts to place values on all of these externalities. The methods for doing so are a major topic in environmental economics and are treated separately in Chapter 11.

Treatment of risk

The future sales to be expected from the investment are forecasts and are subject to risk. If the investment involves more than trivial expenditure then the industrialist will manifest risk aversion, i.e. she will give greater weight to the probability of financial losses from the investment than she gives to gains. A common way of dealing with this problem in investment appraisal is to adjust the sales forecasts downwards to allow for the risk. The individual is averse to financial risk because her resources are limited. Society is not averse to financial risk and would not make any adjustment for it. Individuals can protect themselves against financial risk in two ways:

1. *Risk spreading,* by spreading their investments over a number of projects. Provided the risks attached to these projects are independent, i.e. failure of one investment does not affect the probability of success of the others, then losses on one project will be offset by gains in others. Risk spreading is limited by the individual's financial resources. Society has vastly more resources than an individual and can hence afford to ignore financial risks.

2. *Risk pooling.* The individual can reduce her exposure to risk by pooling her resources with others and sharing with them the expected gains and losses. This is the basis of limited liability, which underlies modern capitalist enterprises owned by thousands of shareholders who share the profits and the losses. Society can be thought of as a giant company with each member as a shareholder. The exposure to financial loss of the individual therefore is trivial and can be ignored.

However society will not be neutral (indifferent) to environmental risks since these can be neither spread nor pooled. The risk of a nuclear accident is an example of an environmental risk. Society cannot reduce this risk by increasing the number of nuclear power stations; that simply increases the risk. Nor can it pool the risk by increasing the number of its members who live near its nuclear power stations; that also increases it. Environmental risks are really negative public goods or public bads. They are non-rival and non-excludable. The individual therefore will free ride but society cannot. Thus the private investor is concerned about financial risk but is indifferent to environmental risk while society is indifferent to financial risk but is concerned with environmental risk.

As with any externality the individual will not be indifferent to environmental risk if it is internalized to her decision process. This would normally be achieved through legal liability. Legal liability in the event of a nuclear accident has been a factor that has prevented the UK Government from privatizing its Magnox nuclear power stations.

Choice of discount rate

The discount rate is the opportunity cost of capital. The social opportunity cost of capital will in general be different from the private opportunity cost and consequently society will use a different interest (discount) rate in making the appraisal.

For the private industrialist we saw that there were two possible sources of finance for the investment project: her own savings, in which case the opportunity cost was the rate of return that she could obtain by lending those savings; or other people's savings, transmitted through the financial system, in which case the opportunity cost was the rate of interest she would have to pay on borrowings. The borrowing rate was higher than the lending rate. Assuming that the costs of what is called financial intermediation, i.e. bringing the funds of lenders to the borrowers which is a real resource cost for society, are covered partly by the borrowers and partly by the lenders, the difference between the rate paid on borrowing and that paid to lenders represents the profits of the financial sector.

From society's viewpoint the profits of the financial sector are a transfer payment that should be netted out and there is thus only one social rate of return on savings. But society differs from the individual in that it has an additional source of investment funding. The individual can only draw on savings, her own and other people's. All private investors are therefore in competition for the available savings and the opportunity cost of one industrialist's investment is investment by another industrialist. The discount rate is the price that allocates the available savings between the potential investors. Society, however, can in addition draw on people's consumption. It can do this through its control of fiscal policy, levying taxes, such as income and sales taxes, thus collecting revenue consumers would have spent on consumption, and spending it on public investment. If the taxes are at the expense of consumption then consumers in effect are being forced to save

Thus there are two possible opportunity costs of public sector investment depending on whether that investment is an alternative to private investment or an addition to it and therefore an alternative to private consumption. If public investment competes for funds with the private sector then the opportunity cost is the (social) rate of return that is earned on private investment. If on the other hand it is funded by forced consumer saving, the opportunity cost is the rate that consumers require on their savings. The social rate of return on private investment is the rate of return that is actually achieved on private investment once adjustment is made for the distorting elements of transfer payments. This rate of discount is generally referred to, perhaps misleadingly, as the *social opportunity cost rate*. The other rate is known as the *social time preference rate*. Technically it is the rate at which consumers discount future consumption.

These two rates of return are illustrated conceptually in Fig. 2.10, which for convenience is reproduced here (Fig. 10.2). The social opportunity cost rate is what was referred to in Chapter 2 as the productivity of investment and is $1-C_{t+1}/C_t$. The social time preference rate is the slope of a contour of the consumer's inter-temporal preference function of which two are shown.

In general it is the case that the opportunity cost rate is greater than the social time preference rate with the differential estimated empirically at 2–3%. This means that investment typically achieves a higher rate of return than people look for on their savings. It was mentioned above that the industrialist will invest provided the rate of return offered is greater than the rate of return that is available on her savings.

We have said that the choice between the two rates for CBA depends on whether the public investment is an alternative to private investment or private consumption. However the opportunity cost of public investment cannot be inferred from the method by which it is financed and indeed the method of finance itself is not normally apparent. Public investment is in the first instance financed by the Government. The Government's two major sources of revenue are taxation receipts and borrowing from the public. An increase in taxation will impact on both consumption and savings. Equally if the Government increases its borrowing, say by increasing the interest rate on national savings,

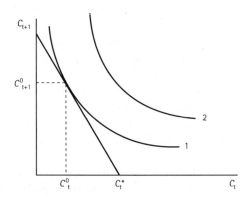

Fig. 10.2 Inter-temporal efficiency in resource allocation.

the public may reduce or divert some of its existing savings or it may increase its savings and hence reduce its consumption. Choice between the two is in fact a matter of macroeconomic doctrine. Older Keynesian doctrines held that public investment was additional to private investment and that therefore the appropriate discount rate was the social time preference rate. Current monetarist views hold the contrary: that an increase in public investment, by pushing up the cost of borrowing, 'crowds out' an equivalent amount of private investment and that therefore the opportunity cost rate is the appropriate one.

In the UK the public sector discount rates are set for each type of public investment by the Treasury. While the Treasury does not give its reasons for the levels chosen it is clear that they are related to the returns achieved in private industry and are therefore closest to opportunity cost rates.

Public goods

If the industrialist cannot capture some of the benefits because of public goods problems (they are non-excludable and the public will free ride), she will ignore them. Society would value them. In practice the issues of valuing public goods and externalities are intertwined since most externalities are compounded by public goods problems. The issues of valuation are discussed in Chapter 11.

Environmental impact assessment

CBA is a tool for public sector decision-making. There is usually no requirement on private sector firms to conduct CBA and we have seen how typical private sector appraisal differs from it. As a means of taking account of environmental impacts of private sector investment, CBA is replaced by environmental impact assessment (EIA). For a range of major projects, an EU directive requires the production of an EIA as part of a planning application so that consideration of the environmental effects of a project are taken into account before a decision is reached (Box 10.1).

EIA differs from the treatment of environmental effects under CBA in two essential ways: there is no requirement under EIA to produce monetary evaluations of environmental impacts and, in consequence, no requirement to formally compare the environmental costs of a project with the non-environmental benefits. An EIA could therefore be a complementary exercise to a CBA providing necessary information on environmental costs and benefits and supplying a formal framework to ensure that all environmental effects receive systematic treatment. In the absence of such a procedure the treatment of environmental effects in CBA has often, in the past, been unsystematic and partial.

The EU directive does not specify whether the projects are public or private investments and many of the mandatory projects would in the past have been public sector projects. Many indeed still would be, but with increasing privati-

Box 10.1 The EU Directive on Environmental Impact Assessment

Under the EU directive environmental assessment has three stages:

1. Applicants for planning permission must submit an environmental statement to the planning authority. This statement identifies the potential direct and indirect effects on human beings, flora and fauna, soil, water, air, climate, and landscape; the interaction between these factors and the effects on material assets and cultural heritage; and the steps envisaged to mitigate or remedy these effects.
2. The planning authority must consult with various public bodies on the statement and also provide the public with an opportunity to express its opinion. To facilitate this a non-technical summary must be included with the statement. The environmental statement must be available to the public.
3. The planning authority must prepare an environmental assessment of the proposal that takes the consultations into account before taking a decision on the planning application.

An EIA is mandatory for a small number of major developments such as power stations and oil refineries, motorways and airports and for some projects that could have major environmental impacts, such as waste incinerators and toxic waste disposal facilities. EIA is also required for projects likely to have significant environmental effects because of their size, location or nature.

For further discussion of the system see Ball and Bell (1991).

---

zation of utilities in all Western countries there will be a growing number of major private sector projects with major environmental impacts. For the UK, as with a number of other EU countries, there is no requirement that such projects should be subjected to a CBA. Environmental economics assumes[3] that CBA will be conducted for all investment projects with environmental impacts, which implicitly therefore will be public sector projects with the environmental effects of private sector investment 'internalized' by the use of economic instruments such as tradable permits for pollution control and the creation of the market for other environmental assets. Since in reality there is a gap, with many environmental impacts of major private sector projects not internalized in this way and with these projects not subject to CBA, the EU directive on EIA fills a necessary role in ensuring that the environmental impact of investment is taken into account in decision-making. The gap could be bridged by a requirement for CBA for major private sector projects. The advantage of doing so, in theory at least, would be that CBA does ask the question of whether the benefits of a project to society are greater than the costs. This is a matter to which I return in Chapter 12. It is beyond the scope of this book to discuss the procedures of EIA in detail.

---

3 And not always implicitly; see Pearce *et al.* (1989).

Summary

- Cost–benefit analysis (CBA) is a technique of investment appraisal for public sector projects. As well as conventional investment projects such as roads, dams and power stations, it can also be used to appraise public policies such as a decision to control acid rain by the issue of tradable permits.
- CBA aims to identify all the social costs and benefits of a project and to appraise the project by a decision rule: invest if the present value of the benefits exceeds the costs.
- No account is taken of the distribution of costs and benefits between individuals and groups in society.
- CBA is used for appraising investment in public goods which, because of non-excludability, would not be carried out in the private sector.
- Environmental effects of private sector investment projects are subject to a requirement for an environmental impact assessment (EIA) under an EU directive. The directive also applies to public sector projects.
- An EIA is a technique for ensuring a systematic consideration of environmental impacts. As such it can complement CBA where the consideration of environmental impacts has often been unsystematic and incomplete.
- EIA differs from CBA in that there is no requirement to place money values on the environmental impacts identified and no requirement to balance environmental costs against project benefits.

CBA differs from private sector investment analysis in the following ways:

- costs and benefits are measured gross of (before) transfer payments such as taxes and subsidies;
- according to the theory of CBA, where prices of factors and goods differ from those that would prevail in a competitive market, 'shadow prices' are used. In practice this is only done, if at all, for certain items such as unemployed labour and imports and exports;
- externalities are taken account of regardless of whether they have been 'internalized' by the use of Pigovian taxes or other instruments. Where possible, external effects should be given monetary valuation so that they can be compared with other non-environmental costs and benefits;
- financial risk is ignored but environmental risks are externalities and are taken account of. This is the exact opposite of the situation with private investment;
- discount rates used for calculating present values are not based on the costs of borrowing or lending but on arguments on what is the opportunity cost to society of the investment. In practice discount rates are specified by the Government.

Chapter 11

Valuing the environment

Since environmental effects, as externalities and public goods/bads, have to be taken account of in cost–benefit analysis (CBA), there is a case for giving them monetary values so that they may be directly incorporated into the framework with other non-environmental effects. Placing monetary valuations on the environmental effects of investment projects is a major activity of neoclassical environmental economics; indeed some prominent environmental economists see it as their primary task:

> The task of environmental economics is . . . to develop existing tools to place valuations on environmental assets and consequences and, thereby, to develop appropriate policies (Helm and Pearce, (1990).

and one, furthermore, that has been subject to much development in recent years:

> For a long period of time . . . it was widely believed that it was difficult, if not impossible, to value public goods or 'bads'. . . . There has, however, been a progressive development during the past two decades, with the result that many goods and services that earlier were classified as intangibles now are classified as measurable (Johansson, 1990).

I disagree with these views and argue that it is not the primary task of environmental economics to place values on environmental assets and that with many environmental assets no meaningful monetary valuations can be derived. Progress has not been made on the matter because, from its nature, progress cannot be made. However it is not necessary to value such assets in order to devise and appraise environmental policies.

To give substance to these criticisms it is necessary to examine the methods of valuation that have been proposed.

An examination of an old text on CBA will yield a wide range of suggested means of placing values on the environmental consequences of investment decisions. None of the suggested techniques has more than a limited range of application and the message conveyed is typically that the analyst should use whatever method would work in the context since any monetary valuation is better than none at all. Modern developments in CBA have rightly led to the abandonment of this *ad hoc* approach in favour of some systematization. *Ad hoc* valuation is indefensible since it can clearly lead to wrong decisions.

For flow externalities the preferred method was, and still is, what can be called *consequential costs. Consequential costs depend on a set of dose–response relationships.* Thus, if a project results in, say, increased air pollution, with rising outputs of SO_2 and particulates, the economist would need to know what were the consequences of that pollution in terms of human health and damage to human artefacts. Thus she would look for estimates of the effects in terms of increased incidence of respiratory diseases and damage to buildings, agricultural production and forestry. These estimates would come from the physical and medical sciences. Given these dose–response relationships she would then estimate the consequential social costs. The human health effects would be evaluated in terms of lost production from workers off sick and having reduced working lives and the increased costs to medical services from the treatment of the diseases. To estimate the costs of damage to buildings it is necessary to know the rate at which stonework decays from acid depositions. The next requirement is an estimate of renovation costs, leading to a final calculation that the additional acid deposition from the increased air pollution would cost society an additional £x per year in maintaining its building stock. Agricultural and forestry damage depend on dose–response relationships between air pollution and (crop and timber) yields; the economist's job then is to estimate the social value of the lost production. The estimation of consequential cost is a legitimate and necessary activity and one to which economists have brought considerable ingenuity. My criticism is not directed at consequential costs.

Consequential cost estimates are an important input in determining pollution standards since it is these impacts on human societies that are of the greatest concern when standards are set. Consequential costs have formed the basis of academic work on global warming. Rising global temperatures are forecast to impact on sea levels and on the location of climatic zones with consequential impacts on agricultural production. The questions then are what costs these effects impose on human populations: in terms of land lost to the sea together with the income it yields, or the increases in coastal defences necessary to protect against rising sea levels; lost world food supplies; population migrations consequent on these changes; increased incidences of tropical diseases from climate amelioration and from population movements. Estimates of these costs are then set against the costs of either avoiding them through policies to reduce the emissions of greenhouse gases, or coping with them.

The consequential cost method depends on the existence of markets that provide the information from which, with knowledge of dose–response relationships, economists can estimate the consequential costs. Where markets do not exist the method cannot be used. Because of this limitation the method in many cases yields only partial estimates of the social costs of projects with important sources of environmental costs omitted. Thus an increased incidence of respiratory disease results in pain and suffering for those afflicted and grief for their families and carers. These are consequential costs of a project that increases air pollution but since there are no markets in pain and grief they cannot be estimated. Grief, pain and suffering are a factor in the appraisal of road schemes since road design can affect the incidence of traffic accidents. The UK Department of Transport allows an (arbitrary) lump sum per accident for 'grief, pain and suffering' in CBA.

Consequential costs also cannot be estimated for the impacts of pollution on wildlife and its habitats, nor for impacts on human society from lost visibility due to air pollution.

External effects of projects for which consequential costs cannot be estimated because of the absence of markets are termed *intangibles*. In addition to some flow externalities, most impact externalities including almost all those that affect the environment are also intangibles and the debate in environmental economics concerns the valuation of these effects. I consider therefore the techniques available for dealing with intangibles.

Expert opinion

One now discredited method of valuation is to seek the opinion of experts in the field. Thus if a project destroys the site of a rare plant, recourse might be made to botanists for a monetary valuation of the loss; if it results in the construction of buildings that intrude on a fine rural landscape, landscape architects might be asked to estimate the cost. The fallacy here is fairly plain. The botanist may understand the significance of the plant loss: knowing how many other sites for the plant exist, how vulnerable it is at those alternative sites, how specialized are its habitat requirements, and so on. She is, in other words, able to provide an appraisal of the value of the site in botanical terms; but this expertise does not extend to monetary valuation. The appropriate expert, if such exists, might be thought to be a person who specializes in digging up rare plants and selling them to unscrupulous collectors! Such a person might have no expertise beyond an ability to identify plants and knowledge of where the rare ones are to be found, but their trade would allow them to produce a monetary valuation. One might be inclined to reply that this is not what was meant at all and that it is the value of the plant *in situ* which is required.

The issue of expertise is one to which we return in the discussion below of contingent valuation.

Cost of replacement

Another long-established method of valuation is to consider what it would cost to replace the asset destroyed or damaged as a consequence of the investment project. Some examples will illustrate the idea. The Commission on the Third London Airport (1971) valued a medieval church, which would have been destroyed if one of the sites it was considering was chosen, at the cost of a new church of comparable capacity designed by a major modern architect. For another site on the Thames estuary, which would have destroyed the feeding grounds of a protected bird species, the dark-bellied Brent goose (*Branta b. ber-*

nicula), it examined the possibility of costing a project to provide alternative feeding grounds by planting the *Zostera* grasses that the birds were eating[1]. The Department of Transport has funded attempts at relocating habitats threatened by road schemes and in other cases attempts to recreate habitats destroyed.

Replacement cost is the cost of action that will neutralize the externality so that if the expenditure is made there is no external effect. The environmental damage of the project is exactly compensated for by the investment so that in that regard society is left no worse off than it would be without the investment. Unfortunately the examples given are all cases of failed replacements. A newly built church, however eminent the architect who designs it, is not the same thing as a medieval church, which has probably no known architect in any event and is likely to comprise a *mélange* of architectural styles and building periods. Its value lies not simply in its religious use, which is presumably compensated for by the new church, but in the historical and architectural interest of the building. It was probably impossible to move the geese and not necessary in any event and the value of the site rested on more than the geese feeding on it. The attempts by the Department of Transport at habitat relocation and re-creation have largely proved to be failures. Only simple habitats have been created, with some replication of the flora but with virtually no success in relocating specialized fauna.

Even if it were possible to exactly compensate for the environmental damage *replacement cost valuations would only be appropriate if the compensation were actually made*. If it is made then there is no environmental damage to value and the cost of avoiding that damage is part of project costs. If the compensation is not made then compensation cost is irrelevant and is no measure of the value of the externality. A non-environmental example illustrates this point. Consider a road-widening scheme that destroys a railway bridge and thereby cuts a railway. If the developers are compelled to replace the bridge then, apart from some disruption during the demolition of the old bridge and building its replacement, the road scheme has imposed no social costs from the loss of railway services and the cost of demolishing and replacing the bridge is properly part of the costs of the road-widening project. But what if the bridge is not replaced? The cost of replacement bears no necessary relationship to the social costs of demolishing the bridge. The railway might carry major intercity traffic whose value far exceeds the cost of the bridge. On the other hand, it might carry one local train a day that hardly anybody uses. The line might even not be used at all.

An environmental example concerns again the Cardiff Bay Barrage. Its construction involved the destruction of tidal mudflats and the loss of feeding grounds for several species of arctic breeding waders. As part of their appraisal

---

[1] This idea was predicated on the assumption that the limiting factor for the birds' population was the availablity of *Zostera* on its wintering grounds. This proved incorrect in a number of respects. *Zostera* species were in fact widespread on the east and south coasts of England and the limiting factors were to be found on the birds' arctic breeding grounds. With the easing of these factors the British wintering population expanded rapidly and the birds began to exploit alternative food sources including the rapidly rising area of winter wheat. Within a very few years MAFF was issuing licences for control by shooting of birds feeding on agricultural land.

of the scheme the consultants to the Cardiff Bay Development Corporation costed the creation of an alternative feeding area constructed by letting the sea into an area of fresh marsh. This cost was used as the value of the environmental resources destroyed by the scheme. Had there been an enforced policy of maintaining the area of feeding grounds for waders then this valuation would have been appropriate, because the costs would have been incurred in the event of the project going ahead and, provided that the calculations were correct, there would have been (from this source) no environmental damage. But there was no enforced policy and, at the time, no intention of creating alternative feeding grounds. Hence the valuation was wrong. What should have been determined was society's valuation of the loss of wader feeding grounds. This might in principle be either less or greater than the costs of replacing them, but in any case is not logically related to it.

The proper way to view replacement cost is in terms of constraints on investment decisions. If the roads authority is constrained to maintain the railway, then the cost of meeting that constraint is part of its project costs. The constraint obviates the need for valuing the external effect since it is designed to prevent that effect occurring or, more accurately, meeting the constraint is deemed to be a sufficient condition for ignoring that particular external effect in CBA.

Environmental effects might easily be dealt with by placing constraints on decision-making and if this is done there is need to place values on the environment. The costs of those constraints are observable from the expenditure that is required in order to meet them, but that expenditure gives no measure of the value of the damage that would have occurred were the constraint not in place. Environmental constraints on investment are probably better described as environmental standards. In the presence of a set of environmental standards there is no need to place values on environmental assets, and standards are therefore an alternative in CBA to techniques for valuing the environment. I return to them anon; meanwhile methods for valuing the environmental effects of investment decisions when decision-making is not constrained by environmental standards are considered.

Surrogate markets

The idea here is that the environmental asset, while not having a market of its own, may be implicitly traded in another market. The classic case concerns a flow externality, noise pollution from an airport, which it is necessary to value in projects to create new runways or terminals at existing airports or to create new airports. If noise is a nuisance that people are willing to pay to avoid, then this should be reflected in the prices of property surrounding airports. Thus if one could observe rented properties identical in all respects except for the level of noise disturbance that they are subjected to, the differential in rents between them would represent consumer valuation of noise nuisance[2].

---

[2] If the properties were owner occupied the house price differential would represent the *present value* of noise nuisance since in purchasing a house the consumer is buying the right to the housing services for an indefinite period.

Noise volume around airports is easily measured and surrounding properties can be classified by contours of a noise index. However, properties will differ in many respects other than noise. It is necessary, therefore, to classify them by the characteristics that influence prices and rentals, such as number of rooms, whether they possess gardens and garages, whether they are detached, semidetached or terraces, are close to shops, public transport and other amenities, etc. If the complete set of characteristics affecting property values could be specified, then a sample of properties differing in their characteristics could be used to estimate statistically the separate influence of each characteristic, including noise nuisance, on house prices or equivalent rentals. Many of the characteristics such as house type and available amenities (presence of garage, garden, etc.) are attributes rather than continuous variables (thus you either have a garage or you do not), which constrains the statistical methodology available. The technique is usually referred to as the construction of hedonic indices. Despite the considerable number of problems involved[3] there have been a number of empirical studies that have provided estimates of the monetary value of noise. The technique has also been used to estimate the effects of a view of a river and of the proximity of a nature reserve on house prices and there are a number of studies from the USA of the impact of air pollution levels on house prices. But it is limited to environmental impacts for which there are surrogate markets and, in practice, to factors that affect house prices.

Travel cost

The standard methodology for measuring the consumer benefits of recreational sites is the so-called Clawson technique (Clawson, 1959). This starts from the proposition that the rational consumer will not incur greater costs to obtain a recreational experience than her valuation of the benefits. This limits the market of specific recreational sites since potential users have to travel in order to get to them and travel, to reach a predetermined destination, is a cost to the user. This cost has two components: the cost of the time spent in travelling and the direct expenditure incurred in travelling (fares on public transport or petrol and wear and tear for private transport). In addition there may be an admission charge for the recreational facility, although the technique was developed to provide estimates of the value of recreation at sites available free of charge and which have the attributes of public goods. The basic idea is illustrated in Fig. 11.1, which shows the demand curve from consumers (household or individuals) for a given recreational site. The assumption is made that the type of use made of the site is the same for all consumers. This may be taken to be informal recreation; if there were a number of identifiable

---

[3] Thus the characteristics have to be independent so that the probability of a house having one characteristic is not predictable from other characteristics (e.g. the probability of having a garage is not greater for large detached houses with gardens). In practice characteristics are typically clustered in this way. More significantly the characteristics of houses in high noise contours differ from those in low contours with the former typically having more amenities, so that the independent effect noise on price cannot be calculated.

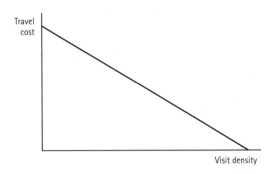

Fig. 11.1 Demand for a recreational site.

different uses (walking, natural history, rock climbing, etc.) then in principle a separate demand would be estimated for each identified use. The information portrayed in Fig. 11.1 would be determined empirically, typically via a questionnaire of visitors determining their frequency of visit, the location from which they came, and the mode of travel. The cost of travel would then typically be determined by the researcher utilizing standard costings for travel time and direct expenditure. Visit density is a product of the number of visitors and their frequency of visits.

If it is assumed that the demand curve for this recreational site is the same for all visitors then the visitor incurring the highest travel costs may be assumed to be marginal in the sense that the value she derives from the visit is just equal to the travel costs she incurs. The highest travel cost is thus a measure of the benefit visitors derive from the site; it is their willingness to pay (WTP) for a recreational visit. Any consumer incurring lower travel costs therefore is receiving a consumer surplus from the visit equal to the difference between the highest observed travel cost, the WTP, and the actual travel cost incurred. For all visitors, the collective consumer surplus is therefore measured by the area under the demand curve, which is taken to be a measure of the consumer benefit from the existence of the site.

The assumption that all consumers have the same demand curve is an unlikely contingency and it is normal to control for variables the researcher believes likely to influence demand. Typical controls might be household income, educational level and demographic variables, such as age of adults and numbers and age of children. An equation of the form:

Visit density = f (travel cost; household income; educational variables; demographic variables)

would be estimated. Where the site is subject to multiple forms of recreational use, information on visit purpose could also be incorporated into the estimating equation.

The Clawson technique is a well tried and tested technique and its problems are well known. It assumes intentionality. Difficulties arise with adventitious visits (people who went out for a drive and just happened on the site) and with visitors who are combining the site visit with other activities (visiting grandma or other sites), since in these cases the travel cost data cannot be used to infer a WTP. The technique also implicitly assumes that if there are several recreational sites the consumer will go to the one with the lowest travel costs. If, by the use of controls, the data cannot be made to conform to that assumption, then the demand curve for the site must differ between respondents.

The technique is only useful for inferring the value of specified sites and only yields estimates of the consumption benefits of those sites. Modern environmental economics recognizes a number of other sources of social value that are not measured by the Clawson technique.

Hypothetical markets

The Clawson and surrogate market techniques only yield information on certain categories of consumption benefits: the Clawson technique gives current recreational benefits where recreation is an unpriced consumption good, and surrogate markets allows the present value of anticipated future consumption benefits to be inferred from house prices. Environmental economics recognizes other categories of benefit that contribute to the social value of environmental assets. The following have been suggested.

- *Option value*. Consumers might be willing to pay to retain the option of consuming an environmental asset even though they have no present intention of doing so. In the presence of option value the value of an asset is understated by market transactions. An example of option value might be a rural bus service that cannot be justified by current levels of use but which rural inhabitants would wish to see maintained as an insurance should their cars break down or they be unable to drive them. Option values are central to the economics of the insurance industry.
- *Existence value*. Consumers might be willing to pay for the knowledge that an environmental asset still exists even though they have no intention of consuming it. Existence value has been cited as the cause of public support for the protection of wild animals and their habitats. Thus it is argued that people wish to save whales from extinction even though they have no expectation, and perhaps even no desire, to see them.
- *Bequest value*. Consumers might be willing to pay to ensure that their descendants have access to the environmental asset. This notion is clearly closely intertwined with that of existence value. If people demand not only that whales exist but that they continue to do so into the future, then they are manifesting a bequest value.
- *Altruistic value*. Consumers might be willing to pay to ensure that the asset is available to others. This is not clearly different from bequest and existence values, both of which are really forms of altruism.
- *Quasi-option value*. This is an obscure term for a fairly complex technical notion. Faced with an irreversible decision about maintaining an asset, the

future benefits of which are uncertain, individuals may attach value to keeping options open. Thus if, on the basis of current information, the optimum decision would result in the elimination of the asset but more information on the value of this asset will become available at some future date, some value may attach to keeping it in being pending the appearance of better information. Quasi-option value forms one plank in the case for the use of the precautionary principle in environmental decision-making, which is discussed in Chapter 15. A number of authors[4] have provided simple two-period examples where there is a positive quasi-option value. A decision has to be taken on whether to develop a parcel of land in period 1 but further information on the value of the land in an undeveloped state in period 2 will not become available until that period. Unfortunately it is also possible to produce counter-examples where the quasi-option value is negative (Ulph and Ulph, 1994). The problem appears to be that the two conditions pull in opposite directions. The prospect of gaining more information at a later date increases the benefits of postponing decisions. However irreversibility as such does not have this effect. Indeed if the value of the asset is expected, on the basis of current information, to decline in the future then irreversibility can lead to an increase in its rate of destruction. This is a widely recognized problem with the exploitation of exhaustible resources such as fossil fuels. The same effect can occur with any exhaustible resource. In a paper for the UK Peatlands Campaign, which aims to protect surviving relics of lowland peat bogs by stopping the commercial extraction of peat for horticulture, I argued that the campaign would have the effect of increasing uncertainty about future prices of peat, introducing the prospect of declining demand, as users switched to alternatives to peat, and of future constraints on extraction. The response of the peat producers was likely to be to increase rather than decrease peat production (Bowers, 1991).

Since quasi-option value remains only a theoretical possibility of uncertain sign, it is not a factor that is taken into account in CBA. I therefore ignore it here.

None of these additional sources of benefit will be reflected in market prices and they cannot therefore be measured by the valuation techniques so far described. However, they can in principle be approached through hypothetical market techniques. There are two basic techniques available: stated preference analysis and contingent valuation analysis. These techniques both rely on sample surveys of members of the public.

Stated preference

Stated preference has been used by transport economists for valuing intangibles in transport. In this technique respondents are presented with a series of choices between options of known monetary value and different quantities or

---

[4] The standard references are Henry (1974) and Arrow and Fisher (1974). The problem is one of dynamic programming and needs mathematics beyond the scope of this text. A less technical exposition – though still reasonably so – is Fisher (1981).

Box 11.1 Revealed preference

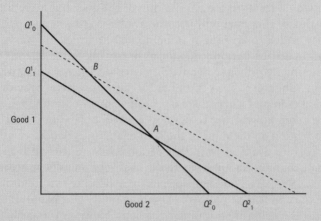

The consumer has a certain money income, which if all spent on good 1 would give her a quantity Q_0^1 and if all spent on good 2 would give her Q_0^2 of good 2. The line $Q_0^1 Q_0^2$ is the outer boundary of her available choice set and its slope reflects the relative prices of the two goods. She chooses point A, which is a combination of the two goods. A is said to be revealed preferred by her. She is now offered a new choice set with the price of good 1 rising relative to good 2 but with her income adjusted so that at the new set of prices she can just buy combination A (the boundary of the new set $Q_1^1 Q_1^2$ intersects that of the old at A). It may be inferred that if the consumer is rational and seeking to maximize her welfare subject to the constraints of the choices available, she will either choose A or some point along $Q_1^1 Q_1^2$ to the right of A. All points to the left have been revealed inferior to A since she could have chosen them in the original situation but preferred A. Since good 2 is the one whose price has fallen she is shown to substitute the cheaper good for the dearer. With many repetitions of this experiment the shape of a preference contour as shown in Fig. 2.5 may be established. The consumer will only choose a point to the left of A if her money income is increased thus increasing her choice set (her consumption opportunities). Thus along the dotted line, which has the same slope as $Q_1^1 Q_1^2$ and hence offers goods at the same relative prices, any point to the right of B could be chosen since these options were not previously available. Revealed preference shows that the assumptions of consumer theory may in principle be determined experimentally.

qualities of the environmental intangible that it is desired to value. By varying the composition of these packages and observing the respondent's choices, her trade-off between marketable goods of known monetary value and the environmental good is determined. The approach is an adaptation of the revealed preference approach to consumer theory (Box 11.1) to the situation where one of the goods does not have a market value. The estimated trade-off is the respondent's WTP for the environmental intangible. Stated preference analysis requires that the intangible can be varied in terms of quantity or quality. It is therefore unsuitable for 'existence' problems where only two extreme states are possible (should this land remain a nature reserve or become the site for a factory?). Environmental effects that are easily scaled, such as noise and visual intrusion, can be valued by this technique; but it is the task of the analyst to present the problem in ways that suit it and, with ingenuity, stated preference analysis can be applied to a wide range of situations.

As a survey-based technique stated preference analysis presents many of the technical problems found with contingent valuation.

Contingent valuation

Contingent valuation is the most widely used technique for obtaining monetary values of environmental problems. It is this technique that is seen as the basis of progress in valuing the environment and which underlies the optimistic quotations given earlier. Over the last decade there have been a large number of studies utilizing this technique and it has been applied to a wide range of problems.

The essence of contingent valuation is very simple. A sample of the population is surveyed and the problem is explained to them. They are told what environmental damage will follow if a project is carried out (the site destroyed and the rare plant lost, the landscape damaged by intrusive development, local air quality reduced, or whatever). They are also told that there will be financial consequences for them if the project is rejected and the *payment vehicle* is specified. What this will be depends on the nature of the problem. Thus if the issue is the construction of a power station it might be suggested that they will face higher electricity prices if it is not built. In other circumstances the payment vehicle might be higher national or local taxes, entry charges or parking fees for a recreational site. One scenario used with site-specific problems is to suggest that the only method of protection is for a charitable trust to buy the site but a fund will need to be created for the purpose. The requirement is that the payment vehicle is comprehensible to the respondents, is perceived as realistic and related to the problem and its proposed solution. The respondents are then asked to indicate their WTP to protect the environment. This is usually done through a bidding game where the interviewer suggests a starting price and raises the price if the respondent accepts it and lowers it if it is rejected. The maximum payment that the respondent is willing to make is her WTP. The value of the environment is then the WTP of the respondents grossed up to the population as a whole.

WTP, as revealed in contingent valuation studies, could encompass all of the sources of value discussed above with the exception of quasi-option value. A number of writers have attempted to divide the total WTP into its various components. A number of techniques have been used for this purpose including providing slightly different scenarios to different subsamples and applying the same questionnaire to a group remote from the problem who, it may be assumed, will have no current consumption possibilities connected with the project and whose valuation may therefore be assumed to be some form of altruism. These sources of value are really motives for valuation and it is not clear that they possess the requirements of separability that allows the total WTP to be allocated into separate components.

I have described in very general terms the most common approach to contingent valuation. An alternative approach is to seek respondents' willingness to accept compensation (WTA) rather than their WTP. For this purpose the scenario has to be designed so that the respondents can be asked what compensation they would require in order to permit the development, with its attendant environmental damage, to take place. The bidding game is then operated in the reverse direction since the WTA is the minimum sum that the respondent would be willing to accept.

In the examination of the Pigovian theory of pollution in Chapter 4 we treated the WTP and the WTA for the individual consumer as identical. Which one applied depended simply on the allocation of property rights. The consumer with the property rights required compensation so her WTA was relevant; for the consumer without property rights it was her WTP that mattered. But in the current context WTP and WTA will not in general be identical since they relate to two different situations. With the former, the consumer is paying to maintain her existing opportunities for benefiting from the environmental asset; in the latter, those opportunities are taken away and her available consumption set is therefore altered. On theoretical grounds, however, WTP and WTA are expected not to differ by much. In practice a number of studies have found substantial discrepancies between the two measures with WTA significantly greater than WTP. This finding has provoked a debate on the reasons for the discrepancy. One suggestion is that consumers are constrained in their WTP by their available income, while no such constraint operates with WTA. The discrepancy would then exist if some people felt very strongly about the environmental issue and, were they richer, would be willing to pay much more to protect the environment. A consumer's WTP for an actual good in a real market is, of course, almost always limited by her income and as consumers get richer they increase their expenditure on most goods. But this income constraint will not normally result in a wide discrepancy between WTP and WTA. A large difference between WTA and WTP would exist for a market good if that good dominated a consumer's expenditure to such an extent that any increase in her income would be spent entirely on that good. The only cases of that sort of problem in practice are probably those of addiction, e.g. gambling, drugs, drink or whatever. Concern with protection of the environment does not seem a likely candidate for addiction. The difference between WTA and WTP is a reflection of the fact that hypothetical markets are not real markets and respondents' behaviour in contingent valuation studies differs in significant respects from market behaviour.

The literature on contingent valuation recognizes a number of sources of bias in the responses to questionnaires. One of the standard works identifies

11 'bias' classes containing 21 sources of bias. Among the major sources of bias identified are the following:

- *Strategic bias.* This arises when the respondent has a strategy with regard to the exercise, which will exist *inter alia* if she has a view on the project. Thus she might be committed to getting it stopped and respond to bids by exaggerating her true WTP (she is even more likely to exaggerate her WTA if that is sought). Strategic bias can also occur if the respondent is indifferent to the project and its environmental impact but not to the interviewer. She may try to please the interviewer by giving the answers that she thinks are wanted; alternatively she may react negatively and try to sabotage the exercise. In either case she has a strategy that will affect the WTP she reveals.

- *Free riding.* This is often included among the class of strategic biases but it is substantially different from the others. If the environmental asset has the nature of a public good the respondent who accepts the scenario and the payment vehicle has an incentive to understate her environmental preference in order to reduce her liability for payment. Free riding will only occur if the respondent perceives that payment is related to her strength of preference but believes that she will be able to free ride on the preferences of others; it is therefore dependent on the specified payment vehicle.

- *Starting point bias.* This will occur if the respondent is influenced in her expressed WTP by the level chosen for the bidding game. This will happen if the respondent has no clear view on her WTP but seeks to learn from the process of the game. This issue is touched on again below.

- *Payment vehicle bias.* The methodology assumes that the respondents are indifferent to the chosen payment vehicle and are concerned only with its costs to them. This may not be true. The respondent may have an aversion to local or national taxes but could be an enthusiastic contributor to charities. If so, she will have a lower WTP for the former than the latter.

- *Information bias.* The respondent may be influenced by the way in which the information is provided and even the order in which elements are given since she may interpret the form of presentation as indicating the relative importance of specific items of the scenario.

- *Hypothetical bias.* The fact that she is not having to operate in a real market where there are penalties for mistakes may influence the care with which she commits her hypothetical resources and the value she attributes to downside implications.

Assessment of contingent valuation (for a more detailed critique, see Bowers, 1993)

The choice of the term 'bias' carries with it the implication that there is a true value of WTP that is the researcher's desire to elicit but which is hidden by the various distortions inherent in the approach or possibly open to correction by careful research design. This is probably the case for some sorts of environmental problems and assets with which the consumer is familiar, such as landscape and tangible pollution (black smoke, noise, offensive smells). But other valuation techniques, such as surrogate markets or travel cost, are also available for these types of problems. *However there is a wide range of environmental problems, including almost all the problems for which contingent valuation is the only technique on offer, where there exists no true value of WTP.* These problems include the valuation of wildlife and wildlife habitats and the 'discovered' pollutions of which the sufferer has no direct perception, which are discussed in Chapter 4.

While there are a wide range of arguments against contingent valuation for these classes of problems, two are decisive.

Information dependency

The respondent to a contingent valuation exercise is supposed to provide her valuation of the problem described. There will only be a true valuation if there exists a true description of the environmental problem to which she can react. There is no such true description and therefore there is no true valuation. A variety of ways of presenting the problem will yield a variety of valuations. For hidden or discovered forms of pollution this point is discussed in Chapter 5 in the critique of the Pigovian theory of pollution. It applies equally in the current context if contingent valuation is used to determine consumer valuation of the impact of a project on global pollution or indeed any form of pollution that sufferers do not directly and independently perceive as a nuisance. I do not repeat the argument here but instead concentrate on the case of valuing wildlife and its habitats.

In many cases there will be uncertainty as to the extent to which a project will damage wildlife habitats. However I bypass that problem and assume that the project will totally eliminate a specific example of a wildlife habitat. The problems that remain then are twofold: the significance of the destruction and the description of the habitat and its associated wildlife.

It is rarely the case that a particular site will contain the only example of a particular plant or animal. A more likely situation is that there will be some species that are restricted in their distribution and are confined to that habitat type. Destruction of the particular habitat then may increase the risk of extinction of those species, but by how much? There is considerable scope for dispute here and opportunity for scientific debate. There is certainly no one unique and indisputable description of the consequences.

But valuation should not depend simply on some of the possible consequences of the destruction of the habitat but on the value of the habitat itself. How is that to be described? Ecologists would concentrate on how the system functions and the interdependence of its component parts and an ecological description of the habitat could be supplied. However, this would be almost certainly incomprehensible to a respondent unless she is trained in ecology. Should she then be given a crash course in ecology so that she can understand the functioning of the system and appraise its significance before she can produce a true valuation of it? One might be inclined to reply that in order to have a valuation, and hence a demand function for a normal marketable good, say a video recorder, a consumer needs to know that it works but not how it works. However, it is as a functioning ecosystem, with its particular interrelationships and components, that the specific habitat is scientifically important.

In order to form her valuation does the respondent need information about the particular plants and animals that comprise it? If so how extensive does that information need to be? Almost certainly the complete assemblage is unknown and will in any case be very large. What sort of summary is required? Should it include microscopic biota, the assemblage of which is likely to be distinctive, or only the things that she could see for herself on a visit? If she is told that it contains a large population of say a particular insect species that is limited in its habitat preferences, does she need to be given a summary of its biology and taught to recognize it? This information would no doubt alter her valuation of the habitat, particularly if the species is beautiful or interesting. Where, in short, is the boundary to be drawn between providing necessary information and assessment by experts? No guidance on this question can be obtained from the theory of consumer behaviour in normal markets since the consumer's reasons for buying the goods is altogether irrelevant; all that matters is her demand, her WTP. It is plainly not irrelevant to her behaviour in hypothetical markets.

There is therefore no one correct or true set of information that can be provided. Since the revealed WTP is dependent on the information set provided there is no true WTP either. The various 'biases' are no more than descriptions of the reactions of the respondent to the dilemma posed.

The valuation mechanism

The thinking underlying contingent valuation seems to be that consumers have a skill in valuing goods and services arising from their activities in proper markets. It is clear that this skill is not innate but has to be learnt. Faced with unfamiliar goods in unfamiliar markets consumers have to learn the market rules and develop a capacity to judge when something is worth buying and when it is not. People entering, for instance, commodity markets for the first time are in grave danger of losing money. A number of aids exist for them, such as consumer magazines analysing products and suggesting good or best buys. The consumer has an incentive to learn since her own resources are put at risk.

There are no markets for the types of environmental assets and problems that are subjected to contingent valuation, (if there were the technique would not be necessary) and consumers have no relevant skills to bring to the hypothetical markets that economists create. In the circumstances, respondents to contingent valuation questionnaires look for clues as to what answers they can give. They try to determine from the description of the problem whether the researchers think it is serious or trivial and take the starting-point bids as an indication of the sort of response that they should offer. They seize upon elements that they can evaluate such as the payment vehicle, since they have experience of paying taxes and electricity charges, and react to that. Alternatively, or additionally, if the problem is one on which they have views, they seek to tailor their responses to achieve their objectives. If they have no views and the problem is one they do not understand, they take their cues from, or react against, the interviewers. The values obtained from contingent valuation are artefacts of the research process and have no independent significance outside of that. There is no important knowledge of valuation that consumers possess and which can be tapped.

Alternatives to valuation

I have argued that, for a wide range of environmental impacts of projects, no meaningful monetary valuations can be produced. Contingent valuation cannot supply them and as such does not fulfil the promises of its proponents[5]. Nonetheless a large number of investment projects do have impacts on the environment and these should be taken into account in decision-making. How then should this be done?

One possible solution, used in the UK, is the application during trunk-roads appraisal of the so-called Leitch guidelines. These involve the construction of a record in standard form of the environmental consequences, which informs a 'political' judgement of whether the environmental losses outweigh the monetary gains. The Leitch guidelines have been criticized for the narrow range of environmental factors that they include but the approach could be extended by, for instance, using an environmental impact assessment as the database for the environmental assessment. A non-monetary assessment of the environmental consequences might serve as a useful filter in project appraisal. Thus if it is possible to determine whether the net environmental consequences are positive or negative and whether they are trivial or serious, we might have the contingency table shown in Table 11.1.

This classification of environmental effects is sufficient to filter out a number of projects where the decision is unambiguous, leaving two classes where there are problems: where the measured net monetary benefits are posi-

Table 11.1 Monetary evaluation

Environmental consequences	Benefit–cost ratio > 1	Benefit–cost ratio < 1
Positive		
Substantial	Invest	?
Trivial	Invest	Do not invest
Negative		
Substantial	?	Do not invest
Trivial	Invest	Do not invest

Note: A benefit–cost ratio > 1 means that the present value of the benefits exceeds the present value of the costs. In the absence of environmental effect this is normally sufficient for a decision to invest.

[5] I am, of course, not denying that monetary valuations of almost anything can be produced by contingent valuation, but simply that these valuations for many classes of effects do not have the meaning assigned to them; they do not form a suitable basis for taking investment decisions that affect the environment.

tive but there are substantial negative environmental consequences; and the converse, where the measured net monetary benefits are negative but there are substantial positive environmental consequences. For these cases political judgement is required.

This approach, however, probably disguises the problem more than it simplifies it. First, the environmental judgement could prove difficult, particularly if there are a number of diverse environmental consequences, some negative and some positive. Even the crude fourfold classification could well involve some difficult and subjective weightings. Second, it seems likely that many projects with serious negative environmental costs will have a benefit–cost ratio >1 so that not much, and certainly not much of consequence, is filtered out. The reasons for expecting this are explained in Chapter 12.

The alternative approach is to protect the environment by the imposition of a set of environmental standards, which must be met in public investment projects. If the standards are met, no environmental damage may be presumed to have occurred. In this case, as explained earlier, the costs of meeting the standards are incorporated in the costs of the projects. A number of consequences of public investment are already protected by the imposition of standards, most notably issues of public safety and health. Thus, in designing a bridge on a trunk road, an engineer does not choose her design level by conducting a CBA to determine the costs and benefits of various degrees of structural stability; she works with pre-specified safety standards. Since environmental standards are central to our view of sustainable development, further discussion of standards is postponed to Chapter 15.

Summary

- Neoclassical economists regard monetary valuation of the environmental consequences of investment projects as a primary task of environmental economics.
- They believe that valuing environmental effects is important for rational decision-making and for implementing sustainable development.
- They argue furthermore that the last decade has seen substantial advances in the capacity of economists to place monetary values on the environment.
- This chapter critically examines that view. It accepts that the measurement of flow externalities by the method of consequential costs based on dose–response relationships is an important function of economics and is central to CBA. Consequential cost estimation is of long standing but of recent years it has been used for larger scale problems such as global climate change.
- The consequential cost method depends on the existence of markets that provide the information from which, with knowledge of dose–response relationships, economists can estimate the consequential costs. Where markets do not exist the method cannot be used.
- Effects of projects for which markets do not exist are called intangibles. These include amenity, human pain and suffering, and impacts on the natural environment and wildlife.
- There are a number of long-established techniques for placing monetary values on intangibles. However they are of limited application, such as inferring the disamenity of noise from variations in house prices and the value of recreational sites from the travel expenditure of visitors.
- Furthermore, these techniques only yield estimates of current user benefits and do not

incorporate non-user benefits, such as existence and bequest value, which are of central concern in the context of sustainable development.

- Hypothetical market approaches are thought to provide measurement of these non-user benefits. The most widely used of these techniques, and the one on which claims of progress are based, is contingent valuation.
- Contingent valuation can be used to provide monetary valuation of many sorts of intangibles including impacts on wildlife and wildlife sites.
- Contingent valuation is a survey approach that asks the public to indicate its WTP for avoiding the intangible consequences of investment projects.
- As a survey technique many sources of bias in responses are recognized. But the notion of bias implies that there exists a true value that can be found once biases are corrected.
- A true value would exist if there was a correct or true description of the problem. There is no true description and hence there is no unique value. The values obtained from the survey depend on the way the problem is presented. Hence contingent valuation cannot provide a correct value of something like a wildlife site; there is no correct value to be found.
- The survey technique assumes that respondents can use their experience in valuing goods in real markets to value intangibles. This is not so. Market behaviour has to be learnt and is specific to the market concerned. Knowing what you are willing to pay for a loaf of bread does not help you to place a value on the existence of blue whales.
- Valuing intangibles is avoided by the use of environmental standards, which serve as constraints on project design. If an environmental standard applies then the cost of meeting that standard is the correct valuation of the intangible for purposes of CBA since it is the cost incurred. These costs are known as replacement or avoidance costs. They are appropriate provided that the expenditure is actually incurred and the environmental standard maintained.

Chapter 12

Cost–benefit analysis in practice

The standard model of cost–benefit analysis (CBA) assumes the existence of a social decision-maker, faced with a series of possible public investment projects between which she is indifferent and seeking guidance as to where the social advantage lies. CBA informs her choice by determining the social costs and benefits for each project, expressing them in a common monetary unit, and summarizing them in the form of a cost–benefit ratio or some alternative such as net present value. It is therefore an aid to rational decision-making on the part of the social decision-maker allowing her to allocate funds so as to maximize the social benefit from public investment.

Environmental economists with this model in mind have been concerned to place monetary values on the environmental consequences of projects in the belief that if this is not done the environment will not receive its proper weighting in the decisions taken. Valuing the environment in this view is thus necessary in order to protect the environment. Extreme versions of this doctrine see any monetary values as being better than none at all, so that concerns about the precise meaning of contingent valuation and other valuation exercises are beside the point. They provide monetary values which ensure that the environment is given some weight in decision-making; without them the environment would be ignored.

The extreme version of the doctrine is clearly untenable. As disscussed in Chapter 10, there are alternative ways of presenting environmental information that do not involve monetary evaluation. The information developed through a properly conducted environmental impact assessment can be detailed and accurate and can be used to make informed judgements about the importance of the environmental impacts. The rational decision-maker would surely prefer this to inaccurate or meaningless statements about the public's WTP. If some of the consequences of public sector projects cannot be expressed in monetary terms then rational decision-making requires that the cost–benefit ratio be supplemented by other decision rules.

But the doctrine fails in its basic premise. The concept of the rational and uncommitted social decision-maker has no counterpart in the actual decision processes for public sector investment appraisal. The case for attaching monetary values to environmental impacts of projects and incorporating these values into cost–benefit ratios can only be appraised from an understanding of how decisions are actually taken and the role that CBA plays in the process.

Reality corresponds to an alternative model to that of the rational social decision-maker. Public sector investment is conducted by specialist agencies. These agencies are not faced with a wide range of possible projects of many types affecting many sectors of the economy. Rather they have a few projects for their particular sector that they wish to carry out. They do not look to CBA to tell them whether these projects are worthwhile; their sectional interest does not allow for that to be questioned. Rather CBA is a strategic tool that enables them to meet their objectives. A satisfactory cost–benefit ratio enables them to obtain permission to proceed with the project and to acquire the funds to do so. It also provides them with a defence against objectors. The logic of the process is then not from cost–benefit appraisal to decision but from decision to cost–benefit appraisal. Agencies are committed to outcomes and the role of CBA has to be judged in that light. I illustrate this process by case studies of two sectors: arterial land drainage and flood protection in England and Wales, and the UK trunk roads programme.

Land drainage and flood protection (based on Bowers, 1988) _____

The administration of land drainage and flood protection in England and Wales[1] has changed substantially with the Water Act 1989. I examine the situation before that Act since this period illustrates clearly the points I wish to make.

Prior to the 1989 Act, responsibility for arterial land drainage and flood protection, including much of coastal flood defences, rested with the land drainage committees of the regional water authorities. Regional water authorities were public sector bodies created under the Water Act 1973 and responsible for sewage disposal and water supply as well as land drainage. In their land drainage activities they were controlled by the Ministry of Agriculture, Fisheries and Food (MAFF), which issued guidelines to them on their duties and supplied grant aid for land drainage works. In addition to ministry grants, land drainage investment was funded from drainage rates levied on land owners and, indirectly through local government, on urban households. Land drainage committees carried out both urban and agricultural drainage and flood protection works. Agricultural drainage was, and indeed still is, also carried out by more ancient bodies called internal drainage boards (IDBs), which exist in a number of wetland areas such as the fens of eastern England, the Somerset Levels, the Kent marshes and the major river valleys of West and East Yorkshire. These bodies operated to broadly similar rules as the land drainage committees of the regional water authorities. I concentrate on agricultural rather than urban land drainage since agricultural drainage provoked substantial opposition from groups concerned with the protection of the natural environment.

Investment in land drainage could take a number of forms depending on the precise nature of the problem identified. At its simplest it could amount to raising the height of sea walls and river banks to protect the land from tidal

[1] The system is different in Scotland; this section is confined to England and Wales.

and fluvial flooding. It might involve deepening and straightening water courses to increase their capacity to move water and thus reduce the risk of overtopping at times of high water flows. Pumping schemes to reduce or eliminate flood waters might also be needed. Finally hydraulic tidal barriers, located on estuaries and raised at times of exceptional tides, could be used to prevent water backing up in tidal rivers and flooding the adjacent land. The schemes promoted aimed to lower water tables on flood-lands and to reduce the risk of flooding

Following the 1973 Act MAFF issued a circular requiring regional water authorities to conduct surveys of their areas to identify all areas of land where flood risk and drainage were unsatisfactory. These surveys formed the basis of works programmes that, over the following 15 years, resulted in threats to almost all of the wetlands of England and Wales and the draining of a substantial number of them.

To qualify for Government grant aid, land drainage investment schemes of both regional water authorities and IDBs had to pass a cost–benefit test. All the schemes were therefore subject to CBA.

The agricultural benefits from drainage and flood protection schemes fall into two categories:

1. *Automatic benefits*: reductions in losses of crops and livestock as a result of the reduced risk of flooding; and increases in crop yields as a result of lowering of the water table.
2. *Induced benefits*: lowered risk of flooding and improvements in the capacity of the arterial system to cope with increased quantities of water allow farmers to invest in field drainage, which enables them to grow higher-yielding and more valuable crops.

Automatic benefits are thus benefits realized without further investment by the farmers. They may involve changes in the agricultural enterprises practised on the land but typically will not. Induced benefits are ones that can *only* be realized by further private investment by farmers. The public investment is the 'trigger' for this private investment. This private investment will normally be in the installation of field drainage systems to provide the drainage necessary for arable crops. Additionally it may involve investment in agricultural machinery and buildings as a result of changes in the balance of farming enterprises from livestock to arable cropping.

Automatic benefits of land drainage schemes are typically very small. This is because agricultural enterprises are adjusted to the nature of the land, and in areas of high flood risk and high water tables are normally no more than seasonal grazing of livestock on permanent pasture. In the event of flooding livestock can swim and can be moved. Standing crops drown and, if the floods are saline, soil fertility is damaged. The composition of grassland subject to flooding comprises species resistant to inundation. These often include rare species, which was one of the reasons for the environmental concerns; others were that these grasslands were breeding and wintering grounds for waders and wildfowl and that they comprised valuable and scarce wetland landscapes.

Thus if farmers owning land on the wetlands were to benefit from the land drainage investment of the drainage authorities, they needed to invest. Most of the benefits from land drainage schemes were induced benefits[2]. The objective of the schemes promoted by the regional water authorities and the IDBs was therefore to bring about reclamation of wetlands by providing the conditions that would induce farmers to install drainage in their fields and then to plough them.

The benefits would then be increased output of arable crops and more intensive livestock enterprises on reseeded grassland. To conduct a CBA of a land drainage or flood protection scheme a drainage authority needed the following information:

- the existing agricultural output of the wetland;
- a forecast of the rates at which it would be reclaimed as a result of drainage works by the farmers (normally termed the 'take-up rate');
- a forecast of the agricultural output from the reclaimed land.

The benefits would be the difference between the forecast output values with the project and what the value of output would have been without it, less any costs of induced investment.

The model underlying the CBA appraisals of land drainage schemes was therefore based on the assumption that the factor preventing reclamation of the wetlands by farmers was the risk of flooding. By reducing the risk of flooding, public sector arterial drainage and flood protection schemes removed that constraint. Some simple algebra will illustrate the farmer's problem.

The farmer has a choice of two enterprises for her wetland. One is unaffected by floods (summer grazing of cattle) and yields an annual income of A. The other (an arable crop) yields a higher income of B if there is no flood, but no income at all if the land floods. So $B > A$, but B requires investment in the installation of field drains, etc.

Let the probability of a flood occurring in any year be p, where $0 < p < 1$.

If the farmer invests, then she forgoes the opportunity of receiving A for the prospect of the higher income B. If there is no flood then her extra income is $B–A$. If a flood occurs she gets no income from the crop so her return from the investment is a loss of $–A$. The probability of receiving this is p. The probability of a flood not occurring (in which case her income is $B–A$) is $1–p$.

Her expected annual income, Y, from the investment is then:

$$Y = (1–p)(B–A) –pA \qquad (12.1)$$

This is the income she will use to calculate the rate of return on her investment in field drainage.

As the probability of a flood falls $(1–p)(B–A)$ increases and pA decreases so the rate of return on the investment in field drainage rises as the flood risk

---

[2] An exception was a pumped drainage scheme for the 'Ings' of the Yorkshire Derwent, a wetland of international importance and protected under several international treaties and EU directives, where the benefits were said to be increased hay yields and an extended grazing season because of the reduction of summer but not winter floods.

falls. If the rate of return is greater than the farmer's opportunity cost of capital r^* then it is worthwhile investing.

The process is illustrated in Fig. 12.1 where the rate of return to the farmer from investing in drainage to produce the arable crop is plotted against the degree of flood risk, p, measured from right to left. Prior to the drainage authority initiating its project, the flood risk is p_0, giving a rate of return on investment in field drainage of r_0. This is less than the opportunity cost of capital r^* so the farmer will not invest. The project lowers the flood risk to p_1. This raises the rate of return on field drainage to r_1 which is greater than the opportunity cost of capital r^*, so it is now worthwhile for the farmer to drain the wetland.

But if this is the model that justifies the investment, it in fact did not hold in practice. One of the most ambitious of the flood protection schemes proposed in the 1980s was the Yare Barrier. This involved the construction of a hydraulic tidal barrier on the mouth of the River Yare at Great Yarmouth, which would have excluded tidal surges from all of the rivers of the Norfolk Broads and resulted, according to the CBA conducted, in the reclamation of all the grazing marshes along the banks of these rivers, and their conversion to growing arable crops. The grazing marshes were already protected from flooding by tidal banks along the rivers. The height of these banks was already such that, even without the Yare Barrier, the flood risk was so low that farmers would have found it worth investing in field drainage and reclaiming the grazing marshes. In terms of Fig. 12.1 flood risk was at p_1 *before* the land drainage scheme. That the farmers had not drained them implied that flood risk was *not* the operative constraint on reclamation and that the assumptions which led to the proposal to construct the Yare Barrier were false. Instead the operational constraint was the fragmented pattern of land-holding, which meant that the consolidation of individual plots into larger holdings was necessary before reclamation was worthwhile.

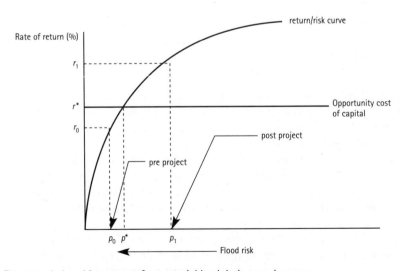

Fig. 12.1 Induced investment from arterial land drainage schemes.

This situation was not confined to the Yare Barrier. As can be seen from equation (12.1) the increased income to the farmer from investing in field drainage and hence the rate of return on that investment depends not only on the flood risk but also on $B–A$, the difference between the income received with and without the investment. As $B–A$ increases the level of flood risk that the farmer is willing to accept increases. $B–A$ depends among other things on the prices for livestock and crops. During the 1970s and early 1980s this difference was large, especially if the crop was a cereal, sugar beet or potatoes. At the prices then prevailing it was worth reclaiming wetlands if the flood frequency was below about one year in five to one year in seven. A large percentage of the wetland areas that were subjected to water authority or IDB drainage schemes had lower flood risks than these levels, so that many of the schemes could not be justified as removing an obstacle to wetland reclamation for arable agriculture.

Indeed pressure from farmers for these arterial drainage schemes existed precisely *because* arable crop prices were high. This was not irrational on their part. Even if reclamation was profitable at current flood risks, reduction in those flood risks would increase its profitability and the farmer would only pay a very small portion of the extra costs of flood protection, the rest being met by the taxpayer in the form of a grant to the drainage authority and from other drainage ratepayers. To demand more flood protection than was needed is an aspect of free riding. Figure 12.1 shows the rate of return to the farmer continuing to rise for levels of p below p^*. It rises at a decreasing rate as the risk of crop loss becomes so low as to become virtually irrelevant to the farmer's decision.

The land drainage and flood protection schemes promoted by the drainage bodies failed to satisfy a basic requirement of CBA, which is that the schemes appraised should be *cost-effective. A scheme is cost-effective if it is the least cost means of meeting the objective.* To determine whether a scheme is cost-effective it is necessary:

- to specify the objective;
- to model the process in order to determine the alternative means of achieving the objective;
- to cost these alternative means of meeting the objective.

Chapter 10 begins with an example of a private investor appraising a potential investment. Her objective was in this case quite clear, to increase her profits, and it was reasonable to assume that she understood how the proposed investment would achieve the objective. It was reasonable also to assume that the investment was cost-effective. Faced with two machines that achieved the same results she would choose the cheaper. If she preferred a more expensive machine this would be because it added more to profits than it added to cost, perhaps because it was more reliable, thereby saving maintenance and lost output from breakdowns.

With public sector investment things are not so simple. The objective is frequently not immediately clear. In the case of land drainage I have inferred that the objective was to increase agricultural output rather than to reduce the area of wetlands for its own sake or to increase the area of higher grade agricultural land (drainage characteristics are one of the factors determining agricultural land grades). I inferred this from the fact that agricultural output was the chosen measure of benefit. The drainage authorities did not try to conduct contingent valuation surveys to determine the public's WTP to eliminate wetlands as they might have been expected to do had the objective been the elimination of wetlands. The same process of inference could be used for the industrialist since her benefit measure is profits.

Furthermore cost-effectiveness cannot be assumed because the mechanism by which the investment achieves its results is unclear and we cannot suppose that alternatives have been considered. These are consequences of sectional specialization in public investment. Since reclamation was prevented by fragmentation of the land, the cost-effective method of achieving the reclamation would presumably be some scheme of grants to encourage consolidation of land-holdings. But the drainage authorities could not consider that because it was not within their powers.

In fact, of course, the drainage authorities' objective *was* to reclaim the wetlands. CBA was a necessary requirement for them to receive funding to carry out this work and they looked for benefits from the investment to justify the funding. The logic is thus precisely the reverse of that of the private industrialist.

Lack of clarity over objectives and failure to achieve cost-effectiveness are not the only problems of public investment by sectional authorities. Once the industrialist has determined the cost-effective method of achieving her objective she still has to appraise it to determine whether it is worthwhile. It may still be that the cheapest machine available does not yield a rate of return sufficient to justify the investment. Chapter 10 demonstrates how social appraisal in CBA differs from the private appraisal of the industrialist. One important difference is that transfer payments should be deducted from the benefits in public investment since they are not part of the social benefits. Transfer payments in the form of subsidies from taxpayers and consumers constitute a substantial component of farm incomes and can amount on occasion to almost the whole of private benefits in agriculture, so the social return from agricultural investment is much lower than the private return. Indeed in the 1990s society is paying farmers not to produce, so the social return from increases in agricultural output is generally negative. Yet, until the practice was changed in the 1980s as a result of sustained protests from environmental groups, the additional agricultural output from land drainage investment was valued by the drainage authorities in their CBA at the prices the farmers received. Even had the schemes been cost-effective and therefore properly subjected to CBA, many of them would not have been profitable had the agricultural benefits been valued net of agricultural subsidies. Despite this many schemes were approved for grant aid and actually carried out.

One final illustration of how the function of CBA is distorted in practice and serves a different function from that assumed in the theory can be given. The industrialist will reject an investment project if its rate of return is less than the opportunity cost of capital. That is not necessarily what happens with public investment. As an example we can look again at the Yare Barrier. Despite being appraised at subsidized agricultural prices, the initial CBA for the Yare Barrier showed the investment to have a negative present value. The response of the land drainage committee was not to reject the proposal but instead to launch a search for additional benefits to raise it to viability. They did this by changing some of the assumptions used in the original appraisal:

- they increased the assumed rate of 'take-up';
- they altered the assumptions about what crops would be grown on the reclaimed marshes towards ones that had higher values[3];
- they revised downwards the value of the existing grazing enterprises;
- they increased the area of land that would be affected by the scheme;
- they decided that the existing river banks were dilapidated and would need replacing in any event, thus making the saving of expenditure on bank replacement a benefit of the barrier.

There are two important points here. The first is that the drainage authority reacted to an adverse CBA by attempting to improve the balance between costs and benefits. This emphasizes the point that the CBA was not conducted in order to decide whether to build the barrier but to justify a decision already taken and to qualify for grant aid. The second important point is that, having decided on this course of action, the authority could succeed. This is because CBA in practice involves making a large number of assumptions about a large number of variables. Altering the outcome of the CBA was possible because of the large number of degrees of freedom that the authority possessed.

The situation described was not that of drainage authorities that were out of control. They were operating according to guidelines issued by their sponsoring ministry, MAFF. MAFF guidelines on the conduct of CBA recommended, with sample calculations, the use of subsidized prices for agricultural output. A MAFF circular told the regional water authorities to survey their areas for drainage 'problems' such as wetlands that they could solve. MAFF also issued guidance on the levels of flood protection that were required for different agricultural enterprises. These were based on scientific evidence of the optimum drainage characteristics for growing various types of crop. They took no account of economics and did not vary with the levels of prices, falling as the prices that farmers received rose to reflect the rising levels of flood risk that the farmers would have been willing to bear.

The fact is that reclamation of wetlands in order to increase agricultural output was ministry policy. This policy was not determined on the basis of CBA but CBA was used to carry it through. The model of the uncommitted decision-maker seeking guidance on how best to allocate social funds does not fit.

The environmental effects of land drainage schemes were not given monetary values in the 1970s and 1980s and in many cases were not considered at all. Would it have made any difference if they had been? Would the wildlife of

---

[3] At the subsidized prices that they were using. The crops were not necessarily higher valued once transfer payments were deducted, but this was irrelevant to them since they were not deducting transfers.

the wetlands and wetland landscapes have been better protected had the drainage authorities been required to put money values on them? Chapter 11 argues that many of the environmental impacts of wetland reclamation, e.g. losses to wildlife, cannot be given meaningful monetary values in any event, but through techniques such as contingent valuation values of some sort could nevertheless have been provided. But this would in all probability have merely generated the sorts of efforts taken in the case of the Yare Barrier to raise agricultural benefits to neutralize them. Given the way the CBA was being used, as a device to justify drainage decisions and to supply the case for grant-aiding them, it is extremely unlikely that a requirement to value wildlife would have been allowed to compromise the objectives.

What might have had a greater impact would have been a requirement to deduct transfer payments, since this would have, and when it was finally introduced did have, the effect of markedly reducing the agricultural benefits, making it difficult to offset them by changing assumptions. A requirement to demonstrate cost-effectiveness might also have had an impact. In short, requiring the agencies to behave like the rational decision-maker might have helped. But this would have required institutional change.

Trunk road appraisal

Trunk roads in Great Britain are defined by Act of Parliament as the most important national through routes. Responsibility for the designation, construction and maintenance of trunk roads in England and Wales rests with the Department of Transport (DoT), while responsibility for other, non-trunk, roads rests with the local authorities. As with land drainage the situation is somewhat different in Scotland and is ignored here. Responsibility for maintenance and construction of trunk roads, including the appraisal of trunk roads schemes, is devolved to an executive agency, the Highways Agency (HA), which is responsible to DoT for carrying out its duties.

While DoT thus has responsibility for the infrastructure of road transport, it does not have similar responsibility for the infrastructure of other transport modes: railway tracks, airports and seaports. In consequence, alternative means of meeting the objectives of long-distance transport of goods and people are not considered in the appraisal of trunk road schemes. The objectives of such schemes are defined solely in terms of road transport (of improving the flow of goods and people by road) and the issue of cost-effectiveness is limited by the institutional framework (this criticism has been made by a number of authorities, most recently by the Royal Commission on Environmental Pollution, 1994). As with land drainage, where the specialist drainage authorities were prevented from defining objectives in the broader terms of increasing agricultural output (which, given the evidence of rising agricultural surpluses, would have been open to question) and had to define their objectives as reducing the incidence of flooding and lowering water tables, so the HA must use the limited objective of improving traffic flows.

The environmental consequences of trunk road construction and improvements are multifarious. Negative ones include:

1. Destruction of wildlife habitats. The 1989 White Paper, *Roads to Prosperity* (Department of Transport, 1989), which doubled the size of the trunk roads programme, would, if carried through, have damaged or destroyed over 160 scheduled sites of special scientific interest (SSSIs).
2. Destruction of archaeological sites. One scheme alone, the Newbury bypass, destroyed 12 sites plus a Civil War battlefield.
3. Intrusion on protected landscapes such as those of national parks and areas of outstanding natural beauty.
4. Increases in local and global air pollution as a result of stimulating road traffic.
5. Pollution of rivers and water courses from run-off of water from roads.

Positive impacts are:

1. Bypasses of villages and urban areas can reduce the impacts of traffic and improve the quality of life for local residents.
2. Reduction in traffic accidents.

The general consensus is that the negative impacts are far more significant than the positive ones.

Trunk road planning goes through a number of stages starting at the choice of a strategic route defined in terms of the origin and destination and in very broad terms the route between them. Many of these stages are internal to the HA and the DoT. While environmental matters are considered at various stages in the process this is mainly in terms of landscape considerations. Other aspects of the environment and CBA are not considered until the late stage of the choice of a specific line for a section of the road. Trunk roads are planned in detail, appraised by CBA, and built in sections typically less than 20 miles in length and often less than 10. These sections presumably have a logic in terms of engineering but none in terms of the economic appraisal or environmental impact.

For any section, a choice of detailed routes is considered and sent out for public consultation before the preferred route is published and, if there are objections (there always are), submitted to a public inquiry, whence the inspector conducting the inquiry submits a recommendation to the DoT for a final decision.

Confining the detailed examination to a small section of a planned long-distance route stultifies argument about the process. It is not permitted, for instance, for objectors to argue that the preferred line for the section will ensure damage in subsequent sections by making it impossible for a particular wildlife or archaeological site to be avoided. Nor can they argue that the road as a whole will stimulate traffic and lead to increases in air pollution and loss of amenity. The inability to argue matters of environmental concern is part of the reason why protest against road schemes has increasingly taken the form of direct action.

Examination by sections, as though the larger plan did not exist, also means that valuing environmental impacts would have little effect on the outcome. Many of those impacts (e.g. air pollution) are attributable to the road as a whole but not to a particular section. Others (destruction of sites further down the line) are consequences of the decision taken but are administratively ruled inadmissible.

Criticism of a road scheme is further curtailed by the treatment of road traffic forecasts. The road traffic forecast is central to the trunk road planning and appraisal process, which has been described as a 'predict and build' approach. DoT produces forecasts of road traffic for up to 30 years ahead. The forecasts are based on assumptions about the rate of growth of national income and a high and a low forecast, derived from high and low assumptions about economic growth rates, is produced. The 1988 forecasts were for traffic to increase over the period to 2025 by 142% (high) or 83% (low).

Box 12.1 Choice of transport mode

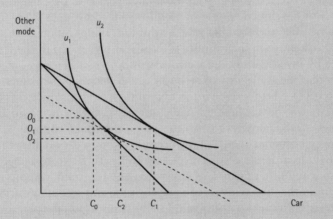

The household's preference function between cars and another transport mode (for simplicity I assume there is only one) is shown in the diagram. The relative prices of the alternatives are given by the slope of the budget lines. Initially the household consumes the quantities O_0 and C_0 of the two modes. The price of the car mode falls because the road system is improved. The slope of the budget line becomes less steep, showing that more car use can be had for the same expenditure; car use has become cheaper. The household is better off because the price of one of the goods it consumes has fallen. It therefore moves to O_1, C_1 on a higher indifference curve. Had it not been better off because, say, the fall in the price of car transport was offset by increased taxes to pay for the road improvements, then it would have found itself on the dotted budget line where it consumes O_2, C_2 more car travel but less of the other mode. The move from O_0, C_0 to O_2, C_2 is termed the substitution effect: it is the change in consumption patterns brought about by changes in relative prices. The move from O_2, C_2 to O_1, C_1 is termed the income effect: it is the change in consumption patterns resulting from the fact that the price fall makes the household better off. The combined effect of the two is the move from O_0, C_0 to O_1, C_1.

Forecast traffic growth is treated as exogenous to the road building programme, i.e. the traffic growth is assumed to be independent of the building of roads, and the function of the trunk roads programme is therefore to provide the capacity for the traffic that will exist. If the roads are not built then the result will be increased congestion and a lengthening of journey times. It follows that new roads do not create traffic, they can only divert it from other roads. Traffic forecasts are treated as government policy and the practice is that objectors may not question them at the public inquiry.

Freight traffic and private car traffic are treated separately in the forecast. Since private cars constitute the bulk of traffic and dominate the growth forecast I confine myself to them. The central plank of the forecast for private traffic is an assumption that car ownership will grow with national income to a saturation level taken to be the level of car ownership in the USA, which has the highest level of car ownership (and the highest level of national income per head) of major developed countries. While the forecast does allow for some increase in the miles travelled per vehicle, this is of minor influence in comparison with the increase in the number of cars. Car ownership in the forecast is not affected by investment in the roads network.

The idea that the availability of road space, and hence journey times, is not a factor in decisions whether to own cars and in the extent to which cars are used by those who own them is fundamentally untenable. People living on an island without roads would not wish to own cars and if they did own them would not drive them. An economic model of the decision to buy a car would see the individual as having a demand for travel and comparing the cost in terms of journey times and other factors, of alternative modes (train, car, bicycle). An improvement in the road system, reducing the time and cost of journeys by car, would increase both car ownership and use for two reasons: people would substitute car trips for trips by other modes; and they would increase their consumption of goods requiring travel (touring, trips to recreational sites, visits to grandma) at the expense of other goods that required less travel (Box 12.1).

The assumption that roads do not generate traffic has been attacked by a number of commentators and was criticized in a recent report by a DoT advisory committee (Department of Transport, 1994). As a consequence DoT has conceded that some road schemes generate traffic and has instructed the HA to modify its appraisal in appropriate cases to allow for traffic generation. However the requirement that the appraisal be confined to a particular section means that it is only traffic generated by the particular section that can be considered. A short section is unlikely to generate more than some local short-distance traffic, such as increasing the use of the car for commuting to work, for which it may make a significant difference in time and convenience. The improvement for the entire strategic route will generate more traffic than the sum of the traffic generated by individual sections, since it will have an impact on longer journeys, including longer distance commuter journeys, as well as short ones. Furthermore the trunk roads programme as a whole will generate traffic that cannot be attributed to individual routes and still less to small sections of those routes. Thus the assumption that road building does not generate long-distance traffic may be a reasonable approximation for the construction or improvement of a short section of an individual road but it is not the section that should be subject to CBA. The project in this case is the entire strategic route and even that is really no more than part of a broader programme for road transport as a whole. It is as though our hypothetical industrialist tried to conduct a separate investment appraisal for each component of her machine. She could not estimate the effects on profits and the exercise would make no sense.

The issue of what it is appropriate to appraise is thrown into relief when we consider the choice of benefits used for the CBA of sections of trunk roads. The appraisal process is based on a computer model known by the acronym of COBA. From a traffic survey giving data on the origin and destination of traffic and a model of the roads network for the area, COBA estimates the time savings for trips that result from the proposed road improvement over the entire period of the traffic forecast. These time savings are then valued. Time savings to commercial traffic are valued at the average wage rate and time savings to private car trips, including commuting to and from work, are valued at a fraction (about one-quarter) of the average wage rate.

For commercial traffic, transport is a cost, and a saving in journey time will, it is assumed, allow companies to save on the wages of drivers. Private motorists will also value reductions in time spent travelling but it is assumed that their opportunity cost of time will be lower than for commercial transport. The argument is as follows. The individual (or the houshold if that is treated as the decision-making unit) will allocate her available time between work and leisure so that, at the margin, the opportunity cost of leisure is equal to the wage rate (i.e she values the last hour of leisure at the wage she could have obtained by working instead). But if the choice is between different uses of leisure, as with leisure driving, the opportunity cost of the leisure hour devoted to travelling to a destination will be lower than this. Additionally some individuals (the unemployed, the retired, and individuals during time when work is not an option such as at weekends and on holiday) will not have the choice of working and their opportunity cost of driving, as opposed to some other forced leisure time activity, will be lower.

But while the choice of the value to attach to time savings, particularly for leisure motoring, is open to argument, the real dispute concerns how these time savings are treated. A section of a road scheme is short and the time savings for the individual motorist on the average journey may amount to no more than a few minutes. Indeed when CBA is used to compare alternative alignments for a particular road section, individual time savings may be a matter of seconds. But the COBA program aggregates these time savings no matter how small. In doing so it treats them as fungible, i.e. that is that a time saving of 1 hour to a single individual has the same value as a saving of 1 minute to each of 60 individuals, or a saving of 1 second to each of 3600 individuals.

While many people would concede that significant savings in journey times have a value to people, the saving of a minute or two would probably not be perceived let alone perceived as valuable. But the fungibility assumption has real impacts on decisions taken. As an example, in the construction of a section of the Knaresborough Bypass in West Yorkshire there was a choice between two routes: one went through an ancient woodland that was an SSSI and would destroy its most important and unusual feature[4]; the alternative, avoiding destruction of the wood, was longer and added 12 seconds to the average journey time. But the predicted traffic volumes on the road section were large, about 17000 vehicles per hour. When aggregated across all vehicles and multiplied by the assumed value of time, the difference in benefits amounted to some millions of pounds and the shorter route, destroying the wood, was therefore chosen.

DoT offer three justifications for fungibility:

1. Small time savings can sometimes result in larger time savings. A classic example is the choice of whether to run down the escalator at a London underground station. If the individual runs down the escalator she boards an earlier train than she would have done had she walked. Catching the earlier underground train enables her to catch an earlier train at a mainline station thereby saving (say) an hour on her total journey time.
2. Individuals behave as though small time savings were valuable to them. They jump traffic lights and pedestrian crossings, reverse out of even quite small traffic jams and follow 'rat runs' with the object of shaving a few minutes off their journey times.
3. Over a long journey small time savings add up to something substantial.

With the first argument, the time saved by running down the escalator is not the few seconds less that it took but an entire hour and should be valued as such. The argument is thus not one for fungibility. There will be very few journeys for which it applies. For the individual without a mainline train to catch, the time saved from running down the escalator is zero.

While the behaviour cited in the second argument is a matter of common observation, it is open to explanations other than that the time saved is valuable. One alternative is that *it is the action of saving time that is valuable and not the time saved itself.* Individuals faced with boring activities such as driving to work relieve the boredom by playing games against nature; seeing how quickly they can complete the journey. The time saved may be worthless to them. Another alternative is that drivers dislike being stationary and will take risks in order to avoid being in a queue or having to stop at lights. In this case they would prefer longer journey times if those journeys guaranteed that they would not face delays.

The third argument takes us to the core of the appraisal problem. The section is only part of a longer strategic route. If a CBA were conducted for the investment in the entire route it would make sense to add up the savings of each section and value the resultant time saving on the entire journey. But that is not what is happening. The CBA is being conducted on the small section as though it were independent of plans for the rest of the route. This is the justifi-

---

[4] An ecotone, the boundary between two different soil types, one acid and one alkaline, where a change in floristic composition occurred.

cation for excluding discussion of environmental effects (air pollution, damage to sites in other sectors) that are outside the sector under consideration. Consistency requires that the benefits as well as the costs be confined to the sector. Within the sector the time savings are small and of little value.

In fact most of the trips reallocated to new sections of roads by the COBA program are very short trips, typically of less than 15 miles. If the time savings on these trips were excluded from the benefits of trunk road appraisal on the grounds that they were too small to be of significance, trunk road construction would fail on a cost–benefit test. What we see here is a prior decision to build trunk roads (the trunk roads programme is a matter of government policy) and a justification of the consequential expenditure by a CBA for each section that is built. The CBA allows DoT to defend its expenditure on the roads programme, and the fungibility assumption, in conjunction with the forecast growth of traffic, ensures in most cases that each sector will be viable and the strategic route will thus be built.

Trunk road appraisal, like land drainage, does not accord to the model of the uncommitted rational decision-maker. CBA plays a strategic role in the implementation of decisions; it does not, and is not intended to, inform the decision-makers whether investment in that sector is an efficient use of society's funds. Given the reality of CBA the environment would not be protected by a requirement that environmental impacts be given monetary values and incorporated into the appraisal. The effective method of protecting the environment is to impose environmental constraints on the decisions taken and require that they meet environmental standards. I discuss this further in the next chapter.

Summary

- The standard economic model of CBA assumes an independent decision-maker, faced with a number of alternative investment projects and seeking guidance as to which one to choose in order to maximize social benefits.
- CBA assists the decision-maker by determining the social costs and benefits of each project, expressing them in a common monetary unit and summarizing them in the form of rates of return or cost–benefit ratios.
- On this view, if environmental effects are not expressed as monetary values and incorporated into the calculus then the environment will not get proper weight in public investment decision-making.
- The practice of CBA does not correspond to this model. Public investment is carried out by sectoral agencies committed to a limited range of projects within their responsibilities who use CBA strategically to achieve their objectives. A satisfactory cost–benefit ratio may be a condition for funding of projects or a means of resisting opponents to schemes. In either case the causation is from decision to CBA rather than from CBA to decision.
- Sectoral agencies are able to use CBA this way because they have control over the appraisal process and are able to define the projects and specify the costs and benefits to achieve their objectives.
- In these circumstances it is unlikely that a requirement to value environmental effects would make any difference to outcomes and therefore serve to protect the environment.
- The argument is illustrated by case studies of arterial land drainage and flood protection of agricultural land and the UK trunk road programme. In both sectors, investment with highly damaging environmental consequences and dubious economic benefits, which would be rejected by the hypothetical independent decision-maker, is approved and carried out.

Part 4

Sustainable development

Chapter 13

Exhaustible resources

I have now explained most of the ideas that are needed for an examination of the neoclassical economist's approach to sustainable development and have presented a critique of them. The important ones are:

- environmental problems as a consequence of market failure;
- the optimum state of the environment (as represented by the optimum degree of pollution);
- the superiority of economic instruments over command and control;
- valuing the environment, i.e. placing monetary values on the environmental consequences of economic decisions;
- cost–benefit analysis as a device for taking public decisions and, with environmental values incorporated into it, as a means of safeguarding the environment.

The final piece to fit into the jigsaw puzzle is resource theory, the theory of exhaustible and renewable resources. These are almost subjects separate from environmental economics for which there are very large and separate literatures. Exhaustible resources fit into the economics of energy since fossil fuels (oil and coal) are exhaustible resources. Renewable resource theory has been developed for the economics of fishing. Nonetheless exhaustible and renewable resources are central to the issue of sustainable development and it is necessary therefore to consider them. Their centrality to the problem of sustainability makes it appropriate to consider them as part of the section on sustainable development. Both exhaustible and renewable resources involve some complicated mathematics. In harmony with the rest of this book I attempt to explain the basic concepts without developing the mathematics that underlie them. However where it is helpful some simple algebra is used.

Conventional neoclassical economics concerns itself with what are known as reproducible resources. Goods and services produced in the economy using the available factors of production can be replicated. Once a consumer good or a capital good wears out, or is broken, it can be replaced. Environmental economics recognizes two other types of resources: exhaustible resources (considered below) and renewable resources (considered in Chapter 14).

Economics of exhaustible resources

Exhaustible resources are primary materials such as minerals and fossil fuels extracted from the earth. Since they are created, if at all, by geological processes with geological time spans, they can be regarded as fixed in quantity, although the total quantity available may not be known.

Because they are fixed in quantity and cannot be reproduced, a decision to extract and use (to consume) a quantity of an exhaustible resource reduces the quantity available for consumption in the future. *Thus the consumption of a quantity of an exhaustible resource at any time carries an opportunity cost, i.e. the value of consuming that resource in the future.* This opportunity cost of consumption is usually given the name *user cost*. The presence of user cost is central to the economics of exhaustible resources. User cost does not exist for conventional reproducible resources since the consumption of an amount now does not reduce the quantity that can be consumed in the future; additional quantities can always be produced.

This feature of exhaustible resources means that the important question for economics is when to use them. Leaving aside issues of recycling, the resource can only be used once. An increase in current rates of extraction and use of coal, oil, metal ores, limestone, etc. reduces the amount that is available for future generations. Economics therefore asks the question of what is the optimum rate of consumption of an exhaustible resource.

The basic economics of exhaustible resources are best demonstrated by considering the simple case of the decision problem of the owner of a mine or an oil-well. Her problem is to determine what her output should be; how much of her asset to extract and sell in the current period. For simplicity I assume just two periods: she either sells the resource today, in period 0, or retains it in the ground until the next period 1. Let the price she can obtain for a unit of the resource today be P_0 and the price she expects to prevail for a unit in the next period be P_1. The cost per unit of extracting the resource and delivering it to the buyer is C, which is not expected to vary between periods 0 and 1.

Because she has a fixed supply of her resource, any sold in period 0 will reduce the quantity she can sell in period 1. If she sells the unit in period 0 she will receive revenue of $P_0 - C$ but forgo revenue of $P_1 - C$ in the following period. The value in period 0 of the revenue forgone is its present value $(P_1 - C)/(1 + r)$ where r is her discount rate. Hence her return from selling a unit today will be:

$$(P_0 - C) - (P_1 - C)/(1 + r)$$

$(P_1 - C)/(1 + r)$ is the opportunity cost of her decision to sell a unit today. It is the user cost of her decision. It arises because she is faced with the alternative of investing her resource to sell it in the following period. Provided that

$$(P_0 - C) > (P_1 - C)/(1 + r)$$

she will be better off selling her resource in the current period. If the inequality is reversed, i.e.

$(P_0-C) < (P_1-C)/(1 + r)$,

she will be better off by investing in her resource by leaving it in the ground. Her *optimum amount of current extraction* is given where:

$$P_0-C = (P_1-C)/(1 + r) \tag{13.1}$$

If the mine owner is to divide her output between the two periods, i.e. if her output in period 0 is not to be either all of the resource or none of it, then provided she is faced with a set of market prices that she cannot affect (a competitive market), the extraction cost C must rise with output. One would normally expect that to be the case.

With reproducible resources there is no element of user cost since resources are produced in each period to satisfy the demand in that period and there is no carry-over from period to period. Hence for a reproducible resource the optimum output is given where $P_0 = C$.

Transposition of equation (13.1) gives:

$$(P_1-C)/(P_0-C) = 1 + r \tag{13.2}$$

This is usually described as the fundamental equation of exhaustible resource extraction. It says that *along the optimum extraction path, where the resource owner is indifferent between the options of extracting or leaving in the ground, the price of the resource, net of extraction costs, has to rise at a rate equal to the discount rate.*

If extraction costs are small relative to the price of the resource, equation (13.2) approximates to

$$P_1/P_0 = 1 + r,$$

i.e. along the optimum path resource prices grow at the discount rate. The interpretation of this result is that the higher the discount rate the faster will the price of the resource rise along the optimum path. A higher discount rate reduces the user cost of the resource and causes mine owners to deplete their resource at a faster rate. Given the level of demand for the resource, the larger supply in the current period depresses current prices and the reduced supply in future periods increases subsequent prices. Thus the rate of growth of prices is increased as a result of supply responses.

The socially optimum rate of depletion is that which is given when prices rise at the social discount rate. If the private discount rate of the mine owners is greater than the social discount rate, then the rate of depletion that the market will deliver will be greater than the social optimum. This is a form of market failure in the markets for exhaustible resources, arising because the private opportunity cost of capital is greater than the social opportunity cost. A control authority might correct this failure by the levying of a tax on production of exhaustible materials.

There are two basic types of tax available to the authority: *ad valorem* and specific taxes (Box 13.1) An *ad valorem* tax will have no effect on depletion rates. Thus if the tax is levied at a rate of $t\%$ the post-tax return to the mine owner is $(1-t)P$ and

$$(1-t)P_1/(1-t)P_0 = P_1/P_0 = 1 + r$$

Box 13.1 Ad valorem and specific taxes

A proportionate or *ad valorem* tax is levied as a percentage of the taxable value. A specific tax is levied as a constant amount regardless of the value of what is being taxed.

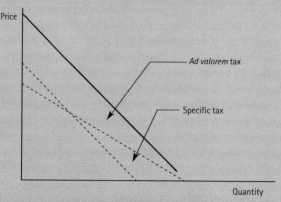

In the diagram the unbroken line represents a consumer's demand for a good. The *ad valorem* tax decreases the slope of the demand curve since the tax take is a constant proportion of the price paid and declines in absolute value as the price falls. The proportionate tax shifts the demand curve down parallel to the original curve since the same sum is deducted per unit consumed and the lower the price paid the greater the percentage of the purchase price that is taken in tax.

Most taxes levied are *ad valorem* taxes. Thus income tax is levied as a percentage of taxable income, although the tax rate varies with different income bands, and value added tax is levied as a percentage of value added. A Pigovian tax, on the other hand, is a specific tax . The sustainability tax considered in Chapter 14 is also a specific tax. Specific taxes are in general the preferred tax type for environmental purposes.

A specific tax will however have the effect of reducing the depletion rate for any discount rate. Thus a specific tax levied at a level $K < P$

$$(P_1 - K)/(P_0 - K) > P_1/P_0$$

and the mine owners will reduce their current rate of depletion until the post-tax ratio of returns, $(P_1 - K)/(P_0 - K)$, is equal to their private discount rate. What in fact they would be doing is to reduce the present value of their tax liability by reducing current production, thereby postponing tax liability to the next period[1].

---

[1] A numerical example may help. Let the mine owners' discount rate be 8.3% and the present price, P_0 be 60. Mine owners will adjust their current production so as to make $P_1 = 65$, since $65/60 = 1.083$. The social discount rate is 7%. The authority levies a specific tax of 10 on the price. Mine owners reduce current production so that P_1 falls to 64.17 since $(64.17 - 10)/(60 - 10) = 1.083$. The ratio of prices before tax, $P_1/P_0 = 64.17/60 = 1.07$. The specific tax reduces the private optimum depletion path to the social path.

A specific tax on an exhaustible resource is usually known as a *royalty*, royalties are levied by many governments on the production of exhaustible resources. If the government wishes to capture some of the proceeds of the production of the resource for itself or its citizens without altering the rate of exploitation then it would levy an *ad valorem* tax.

Resource scarcity

The analysis so far predicts rising real prices as stocks of exhaustible resources are used up as a consequence of decisions by resource owners on extraction rates. This analysis paid little attention to extraction costs and in the simplified version they were assumed away. In fact it is expected that extraction costs would increase over time as a result of what is known as Ricardian scarcity. This doctrine, first developed by the nineteenth-century economist David Ricardo, holds that the most productive deposits will be the first to be exploited and that, through time, the costs of production will increase as less rich and more expensive deposits are exploited. Ricardian scarcity could operate within the individual mine; the mine owner will exploit the reserves that are easiest of access and the cheapest to extract first and will thus postpone (and therefore discount) the investment necessary for exploitation of the more difficult sections, and between sources. The richest mines and oil-wells thus are exploited first with the exploitation of more expensive sources coming on stream as the price rises and the more profitable sources are exhausted.

The predictions of Ricardian scarcity were examined empirically for the USA in a famous study by Barnett and Morse (1963) and updated and extended to the rest of the world subsequently by Barnet (1979). The authors considered two versions of the scarcity hypothesis: a strong version, that unit costs of extractive (exhaustible material) industries should rise through time; and a weak version, recognizing the impact of technical progress in lowering costs of production in modern economies, that costs in production industries should rise relative to non-extractive industries.

Extractive industries were defined broadly by Barnett and Morse to include not simply minerals and fossil fuels but also agriculture and forestry, which utilize the exhaustible resources of the soil. The initial study was for the period 1870–1957. Barnett's update extended the period to 1972. The findings were unambiguous. The strong version of scarcity was totally rejected: all extractive industries experienced continuously falling costs. The weak hypothesis also received no support except in forestry. Forestry is arguably not an extractive industry at all in the true sense, since much of the product is derived from natural forests and therefore falls in the category of renewable resources, examined in Chapter 14. For all true extractive industries, costs over the long term fall relative to costs in non-extractive industries.

These studies also examined the proposition, derived in the previous section, that because of the influence of user cost, prices of exhaustible resources should rise relative to those of reproducible resources. This proposition is also decisively rejected by the evidence. While the prices of many exhaustible materials are subject to considerable fluctuations and are in general considerably less stable than those of manufactured goods, there is no evidence that they are subject to a long-term upward trend. The exception again is that of forestry products, although the last 20 years cast doubt on even that as an exception.

Kerry Smith (1979) suggests four basic reasons for the failure of the scarcity hypothesis:

1. As higher-grade sources are exhausted, lower-grade sources are found in greater abundance. Furthermore the difference in grades diminishes as the known stock expands.
2. As a particular resource becomes scarce, price rises are offset by switches in demand to substitutes. That is, scarcity is offset by declines in demand.
3. Increases in prices stimulate exploration for new deposits and provide pressures for increased recycling.
4. Technical progress acts on supply by reducing extraction costs and by making possible the exploitation of previously uneconomic deposits. It also reduces demand by encouraging efficiency in resource use.

Exhaustible resources and the limits to growth

Despite the lack of evidence of scarcity, the likelihood of a shortage of exhaustible materials providing a limit to economic growth was an issue of controversy in the 1970s following the publication of the Meadows Report (Meadows *et al.*, 1972). The Meadows Report approach was very simple and ignored completely the issue of economics (the real concern of the report was not income growth but population growth). The authors assumed that exhaustible resources were available in fixed quantities and that the demand for them was subject to exponential growth. These assumptions led them to predict that the world would run out of various materials at specific dates and to express available known reserves in terms of years of life left. These ranged from 111 years for coal down to 20 years for petroleum and 13 years for silver and mercury. Even if known reserves were increased by a factor of five to allow for new discoveries, the assumption of exponential growth in demand meant that the effective lives of reserves were not increased in proportion.

In Ricardian economics resources are abundant but have declining quality and the highest grades are used first. In the Meadows Report resources are fixed in quantity but constant in quality. Furthermore the price mechanism does not serve to ration them through time. This position is plainly untenable and estimates of reserves of many materials today substantially exceed even the higher estimates of the Meadows Report.

This issue of resource scarcity has declined in importance in recent years for several reasons:

• Concern has switched to the environmental consequences of exhaustible material exploitation and, particularly with fossil fuels, to the implications for regional and global pollution resulting from their use.

- The biosphere is seen as a more likely candidate than the supply of minerals and fossil fuels as the ultimate constraint on economic growth.
- The methodology of the Meadows Report and the predictions derived from it were so defective as to undermine the message.

To these reasons might be added a fourth, namely that exhaustible resource theory has become a specialized discipline in its own right with concerns that are increasingly divergent from the central issues of environmental economics and which can be given no more than cursory treatment in this book.

The importance of the Meadows Report lies not in its doctrines but in the fact that it was the first authoritative source to raise the possibility that the earth was finite and that its capacity to accommodate human economic activity was subject to limits. As such it is the direct ancestor of current concerns with sustainable development, which is the subject of the next chapter.

Sustainability and exhaustible resources

The problem that exhaustible resources pose for sustainability is that they are exhaustible. Although recycling and reuse, depending on technology, is possible to a limited degree, use at one period denies their use in the future. Thus it is not possible for current generations to leave future generations the same quantities of exhaustible resources that they themselves have access to. The notion of sustainability therefore requires interpretation. It can be interpreted as requiring improvements in efficiency of extraction and use so as to maintain constant the economically effective level of stocks. This condition is satisfied if user cost is constant over time.

This condition can be met by technical progress along several dimensions:

- improvements in efficiency in primary use;
- efficiency in recycling;
- efficiency in extraction;
- replacement of the exhaustible materials by renewable alternatives.

The possibilities are illustrated in the energy sector, e.g. improvements in efficiency of use for particular fossil fuels, the substitution of fuels with greater thermal efficiency for those of lower efficiency (natural gas for coal), the development of renewable alternatives (wind, wave and solar power), techniques for the use of waste products (methane from landfill sites), technology to exploit stocks previously unavailable (oil from under the sea and from the polar regions), and techniques for energy conservation in consumption. Sustainability requires that, in sum, these improvements should leave successive generations facing no greater constraints on the availability of energy than are experienced by the current generation.

Under a sustainability strategy user cost would fall, in part as a result of falls in extraction cost but also as a result of price falls as alternative reproducible products, which do not have user costs, satisfy demand for exhaustible resources.

The first three alternatives listed above extend the use of exhaustible resources, increasing their expective life. They therefore buy time for the final alternative, which some commentators would see as the only true option for sustainable development. While this view may be logically correct, the amount of time that is bought by *continuous* improvements in extraction efficiency, in recycling and efficiency of resource use could be very long. We have seen that there is no evidence of rising material shortages despite decades of unrestricted economic growth. In contradistinction to the pessimists, as represented in the Meadows Report, are the cornucopians. Economists of this persuasion see the process of economic growth as generating the technological improvements that prevent the emergence of raw material shortages.

Summary

- Economics recognizes three types of resources: reproducible resources, exhaustible resources and renewable resources. Conventional economics is concerned with reproducible resources.
- Exhaustible resources are available in a fixed finite quantity, although what that quantity is may not be known. Additional quantities cannot be produced. Exhaustible resources include fossil fuels and metals and minerals that are mined.
- The essential feature of exhaustible resources is that use at any time precludes use at a future time. Hence any decision to use entails an opportunity cost in terms of future options for use forgone. This cost is known as user cost.
- User cost is the present value of future revenue that is forgone.
- Decisions on current use depend on a comparison between current revenue and user cost. The optimum rate of extraction/use is given where the current revenue from the extraction of a unit of the resource is equal to its user cost.
- Along the optimum path of exploitation the price of the resource will rise at the discount rate used by resource owners.
- If the social discount rate is less than discount rate used by private owners of exhaustible resources, then the rate of exploitation of resources will be greater than the social optimum rate.
- A control authority may reduce the exploitation rate by imposing a specific tax, usually known as a royalty, on resource producers.
- Resource theory predicts that the price of exhaustible resources should rise relative to the prices of normal reproducible goods but there is no evidence that this is happening.
- The predictions of the Meadows Report, the originator of the sustainability debate, were that the world would rapidly run out of exhaustible resources and that this factor applied limits to growth. This has not happened and recent discussions of sustainable development place more emphasis on the capacity of the global atmosphere and the biosphere to sustain human life and less on the problem of scarcity of material resources.
- From the nature of exhaustible resources it is not possible to leave future generations with the same quantities that are possessed by current generations. Sustainable development is therefore interpreted as requiring technical progress in use and extraction and the development of renewable or reproducible resources that fulfil the same functions as exhaustible resources.

Chapter 14

Renewable resources

Renewable resources are natural plants and animals that are exploited by humans. The notion excludes plants and animals that are domesticated and whose populations are subject therefore to direct human control. Thus renewable resource theory does not apply to fish farming or agriculture. It applies to fisheries other than fish farming and to any animal that is hunted. It would also apply to the collection of wild plants.

Renewable resource economics

Like exhaustible resources, renewable resources cannot be reproduced. Available quantities are only indirectly affected by human populations via their levels of exploitation. The economics of renewable resources therefore depends on assumptions about their population dynamics.

Figure 14.1 shows the population dynamics of a hypothetical renewable resource, which I assume to be a species of edible fish. The solid curve relates the population growth rate of the species, measured as the net annual increment to the population (births minus deaths), to the population level. N_{max} is the population level that will obtain as a result of natural forces in the absence of human exploitation. At N_{max} annual births equal deaths from disease, predators and ageing processes and the population is just replacing itself. If the species is subject to a negative shock that reduces the population below N_{max} then it will recover because for all population levels below N_{max} growth rates are positive and population growth will ultimately bring the population back to N_{max}. Thus if a natural disaster, say an undersea volcanic eruption, reduces the population to N_2 the population will increase in the first year by Y_1. In the next year the population level of $N_2 + Y_1$ will grow by less than Y_1 but the growth rate will still be positive and the process will continue until the population is back to N_{max}. The time path of the recovery cannot be read from Fig. 14.1 but may be readily calculated from it.

A volcanic eruption is a one-off shock. Predation of the population is a continuous process. The arrival of a new predator into the environment will have the effect of reducing N_{max} and the population will stabilize at a new level, thus accommodating the predator. So if a predatory fish colonizes the species' living space and consumes Y_1 fish per annum, the effect is to shift the horizontal axis up to the dotted line: N_{max} becomes N_2.

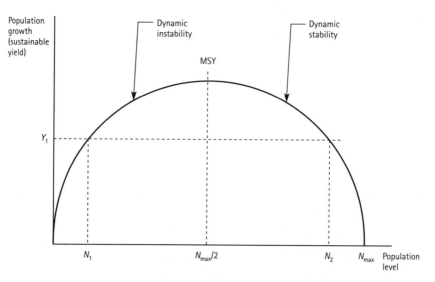

Fig. 14.1 A sustainable yield curve. MSY, maximum sustainable yield.

Artificial disasters such as pollution, which affect the biology of the species by reducing its life expectancy through long-term poisoning, will have similar effects to the introduction of a new predator. However if pollution affects the species' ability to breed, as happened to European birds of prey in the 1960s through the build up of dieldrin pesticide residues in body fat, the shape of the whole curve will be altered and recovery speeds reduced. Loss of habitat from human agency is likely to similarly modify population dynamics.

At population level $N_{max}/2$ population growth rate reaches a maximum known as the maximum sustainable yield (MSY)[1]. *To the right of the MSY the population is dynamically stable.* A population reduction increases the growth rate and, in response to additional shocks or the introduction of further predators into the system, the population falls and the growth rate rises to accommodate.

To the left of the MSY the population is dynamically unstable. A further one-off shock will reduce the growth rate and the population will be unable to accommodate additional predators. Thus if a natural disaster reduces the population level to N_1 its growth rate will rise to Y_1, the same level as if the shock had been less traumatic and had reduced the population only to N_2. The population will still recover and will ultimately return to N_{max}, but with the difference that a further shock, reducing the population below N_1, will reduce, not increase, the growth rate. If at N_1 a new predator consuming Y_1 per annum arrives then the population can accommodate it and will remain at N_1. But it will be unable to accommodate any further predators; if others enter the envi-

---

[1] The population growth curve may not be symmetric and the MSY may thus not be at $N_{max}/2$. It is conventional to show curves as symmetric partly because the equation to such a curve is simple and readily manipulated mathematically. For our purposes the precise shape of the curve is not important and is in any case rarely known.

ronment then the take of the initial predator must fall below Y_1 since that is the maximum predation that the system can accommodate.

Human predation

If a fisherwoman decides to exploit our hypothetical fish species then the population growth curve tells her what sustainable yields are available from the fishery. Thus if she chooses a yield of Y_1 per annum the population will settle at N_2 and she can continue to catch at this rate for ever. From the viewpoint of the fish population the fisherwoman is simply another predator. However there are two population levels that will yield Y_1, N_1 and N_2, differing only in that one is dynamically stable and the other dynamically unstable. Her choice between them is determined by economic factors and there are reasons for expecting that she will end up preferring the unstable position of N_1.

Assume that the fisherwoman plans to sell her catch The species is seen by consumers as a close substitute for other fish and its contribution to total fish supply is small. In these circumstances the fisherwoman has to accept the market price and can sell as much as she wishes at that price. In effect, while being a monopoly supplier of this particular fish species, she is faced with perfect competition. The situation is equivalent to that of the owner of the single mine in the last chapter. To strengthen the similarity further I assume for the present that she has secure property rights in the fishery. No one else can fish for this species.

If, for reasons yet to be determined, she decides that in the long run she will supply Y_1 to the market she has two possible strategies:

1. She can supply Y_1 from the beginning. The population will fall to N_2.
2. She can catch more than the MSY for a period until the population falls to N_1 and thereafter catch Y_1.

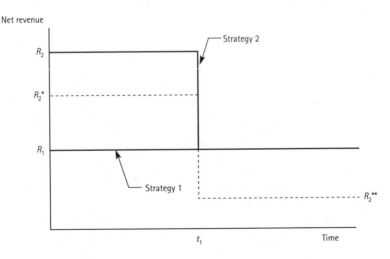

Fig. 14.2 Alternative fishing strategies.

The revenue streams from these alternatives are shown in Fig. 14.2. Strategy 1 gives her a revenue level of R_1 for the indefinite future. Strategy 2 gives her a higher revenue of R_2 for a period until t_1, when the population level falls to N_1, and thereafter revenue R_1.

It is clear that she would be better off choosing strategy 2. It has a higher present value to her for any positive discount rate. The only argument against strategy 2 is that it carries greater risk in that any negative shock to the fish population or the entry of any new predator will reduce the population below N_1 and her sustainable yield below Y_1. If she recognizes these risks she may well choose to ignore them, calculating that the higher returns in the short run more than compensate for the longer-term risk of lower returns, or reasoning that she can always respond by reducing her catch for a period to allow the population to recover. But the chances are that she will ignore these factors altogether, treating them as people treat the risks of being hit by a meteorite or having their houses destroyed by earthquakes: as part of the inescapable environmental background.

There are, however, two economic factors that may lead the fisherwoman to choose the the dynamically stable strategy 1.

The first arises if we modify the assumption about the nature of the market and assume that the fish species is not a perfect substitute for others. In this event the demand for catch will be less than perfectly elastic and the higher catch in the period until t_1 could only be sold at a lower price. This possibility is shown by $R_2{}^*$ in Fig. 14.2. The less elastic the demand the greater the fall in price and the closer $R_2{}^*$ would be to R_1. If demand elasticity were < 1 then $R_2{}^*$ would lie below R_1. However this is unlikely and with $R_2{}^*$ as shown strategy 2 remains the dominant strategy, being preferred at any positive discount rate.

The second requires the introduction of the cost of fish catches, which I have so far not mentioned. Figure 14.3 portrays what may reasonably be assumed about cost curves expressed as average costs per unit of fish landed for different quantities landed.

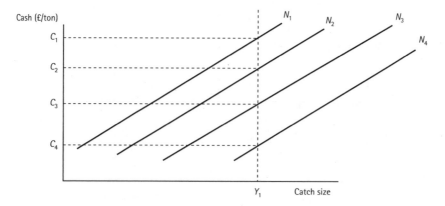

Fig. 14.3 Average cost curves.

Cost curves are shown for four levels of the fish population with N_1 the lowest and N_4 the highest. The curves show unit cost rising as catch size increases as a result of increasing effort needed (longer hours of search, etc.) with the given capital equipment (boats, nets, etc.). Whether the fishery has decreasing returns to scale, i.e. rising costs in the longer term when capital equipment levels are varied, does not matter. The important thing is that unit costs for any catch level are negatively related to the size of the fish population Thus for yield Y_1 costs rise from C_4, for the highest population level, to C_1 for the lowest level. The reason for this is straightforward: as the population declines more effort has to be expended to locate and catch a given volume of fish. At N_4 there are plenty of fish so that relatively little effort is needed to obtain a given yield. At N_1 fish are scarce and a lot of time has to be spent in finding them.

The consequence of this is that costs will be higher under strategy 2 than strategy 1. They will initially be higher because the fisherwoman is further up her cost curve as she is making a heavier catch. But because her catch is unsustainable, the population will decline and she will shift to higher cost curves. After t_1 she will be on cost curve C_1 with strategy 2 and will receive lower revenue from Y_1 than she would under strategy 1 where the population level remains high. The revenue under strategy 2 is therefore given by the dotted line in Fig. 14.2, starting at R_2^* and falling to R_2^{**}. Strategy 2 is then no longer the dominant strategy at all discount rates. At high discount rates the fact that $R_2^* > R_1$ will make strategy 2 preferred, but at lower discount rates the fact that, after t_1, $R_2^{**} < R_1$ enters the calculation.

There will be some discount rate at which the gains in the period before t_1 exactly balance the losses in the period after t_1. Call this discount rate r^*, which is the internal rate of return on an investment project for the fisherwoman to switch from strategy 1 to strategy 2. Then she will choose the lower, dynamically unstable, equilibrium if her opportunity cost of capital $r > r^*$.

The argument may be summarized as follows[2]: *the probability of the population of an exploited fishery being reduced to dynamically unstable levels will be higher:*

- *the higher the discount rate;*
- *the more elastic the demand for the product;*
- *the shallower the slope of the cost curve with respect to yields;*
- *the less costs increase as population levels of the prey species fall.*

Optimum exploitation path

The analysis has proceeded on the assumption that the fisherwoman had decided in advance on the long-term yield that she wanted. That was an arbitrary restriction made in order to simplify the argument. The economics

---

[2] For those familiar with calculus the statement can be expressed as follows. If $_cF^Y$, $_cF^N$ are the partial derivatives of unit cost with respect to yields and population levels, e_d is the elasticity of demand for the landed catch, and r is the fisherwoman's discount rate then the probability that $N < N/2$, $p(N < N/2)$ varies positively with $_cF^Y$, $_cF^N$ and r and negatively with e_d.

of renewable resources concerns itself with the optimum exploitation path. *The optimum exploitation path is the exploitation path that maximizes the present value of the fishery to the fisherwoman.* Given her discount rate, she will choose from amongst all the available exploitation strategies the strategy that gives the highest present value of income from the fishery. We can think of this as a two-stage process involving, first, the choice of yield that under strategy 1 produces the highest annual income and, second, examining whether there are type 2 strategies that have higher present values than the best type 1 strategy.

Strategy 1 may be applied to any constant positive yield up to and including the MSY and all will leave the population at a dynamically stable level between N_{max} and $N_{max}/2$ (see Fig. 14.1). The MSY is unlikely to be the most profitable yield: it will have higher costs than any other sustainable yield having both the highest yield and the lowest population level ($N_{max}/2$ is the lowest dynamically stable population), and if the elasticity of demand is less than infinite the catch will command the lowest price of any strategy 1 yield. Assume that the best yield is Y^*, i.e. $0 < Y^* <$ MSY. Given Y^*, a range of possible strategy 2 options exist. Figure 14.4 illustrates the process. The highest strategy 1 yield Y^* gives a constant net revenue of R^*. There are a large number of possible type 2 strategies and I illustrate three. Type 2 strategies involve a period of exploitation above the MSY followed by exploitation at a constant rate. Assume that the rate of exploitation above MSY is the same for all options and is set so as to maximize revenue net of costs. The options differ in the period t for which yields greater than MSY are taken; the longer is t the lower the resulting population and the lower the subsequent yield. Thus Y_1 has the shortest period and consequently the highest subsequent yield and revenue R_1; Y_2 is intermediate and Y_3 entails the longest period of high yields with consequently the lowest sustainable yield and revenue R_3. At any discount rate one of these will be the best. At low discount rates the high penalty imposed in the longer term by Y_3

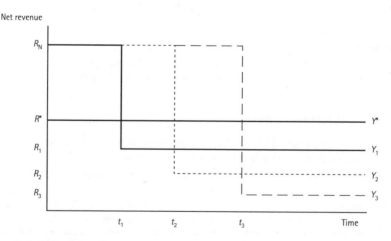

Fig. 14.4 Strategy 2 options.

may make Y_1 or Y_2 the preferred option. The higher the discount rate, the longer the period t that is preferred and the lower the subsequent dynamically unstable yield that results. Viewed this way, except for very unlikely values of demand elasticity and parameters of the cost curves, there will always be a type 2 strategy that dominates a type 1 strategy. Hence the probability of populations of renewable resources being depleted to dynamically unstable levels is very high.

Common access problems

This conclusion is reinforced once we abandon the assumption that there is only one person or firm exploiting the resource and that entry into the market by others is somehow prevented. The reality for many renewable resources and particularly for fisheries is that of open access. Open access exists when anyone, or any of a large number of people, is able to enter the industry and exploit the stock. Open-access problems exist for fishing in international waters in the absence of international agreement on fishing limits. But it also exists in territorial waters unless there are restrictions on who can fish and how much fish they may take.

Open access leads to the *tragedy of the commons*. With open access, conservation strategies by our fisherwoman are ruled out since any attempt she may make to save fish for subsequent periods could easily be neutralized by others entering her market and depleting the resource. The tragedy of the commons exists when there are no defined property rights in renewable resources. Strategies 1 and 2 are ruled out and existing fishermen and women will simply set their yields so as to maximize current revenue. With open access, if economic factors lead to over-exploitation, the only factor that can prevent the species being driven to extinction is the rise in costs resulting from declining stocks, which may make fishing uneconomic, while the species still has capacity to recover. Even so, rises in stocks will cause those who have left the market to re-enter it again so that the population fluctuates at low levels and the species is always at risk from technical progress in fishing methods making fishing profitable at low population levels.

The traditional solution to common-access problems is for states to take the property rights in fisheries and to control entry to them or to control catch levels by imposing quotas backed by the law. Recently alternative marketable instruments have been tried. These include transferable or marketable quotas, designed to overcome the supposed allocative inefficiencies of command and control, and property-rights systems whereby fishermen and women are given entitlements to a certain share in whatever is agreed as the permitted catch. These systems seek to overcome the tragedy of the commons by guaranteeing participants a share in whatever benefits accrue from conservation of stocks (for a discussion of property right approaches see Young, 1992, Ch. 4).

Sustainability and renewable resources _____

Long-term survival of renewable resources requires that populations are kept at levels where yields are dynamically stable; in other words that the population levels be kept above those of MSYs. This is necessary to provide some protection against unanticipated shocks to the population: new sources of predation, natural and man-made disasters and disease. With many renewable resources current population levels are at dynamically unstable levels and the population faces risks of extinction that are incompatible with sustainability. This is the case with many pelagic fish: haddock and cod and notably the North Sea herring.

One strategy to increase the yields of renewable resources without endangering stock levels is to farm them. Farmed species are not renewable resources since the reproduction of the species is then within human control. The farming of salmon, trout, carp and other species allows high levels of consumption without endangering wild stocks. Deer farming and crocodile farming achieve the same ends for those species. With plants the commercial growing of edible fungi offers a means of saving European fungi. However, farming depends on the ability to control reproduction. With the more valuable species of edible fungi, notably cep (*Boletus edulis*), chanterelle (*Cantharellus cibarius*), morel (*Morchella vulgaris*) and truffle (*Tuber aestivum*), this is currently not possible. Farming, as opposed to exploitation of wild stocks, is dependent on technological development. Investing in research on techniques of cultivation of threatened renewable resources can form part of a sustainability strategy for renewable resources.

Renewable resources in practice _____

The economics of renewable resources is dependent on the population dynamics of the target species. These dynamics determine the shape of the sustainable yield curve for the species. I have made the very simplest assumptions about population dynamics and these are incorporated in the sustainable yield curve of Fig. 14.1. They were made in order to derive some broad conclusions about renewable resources that inform the discussion of sustainable development. In the discussion the population level of the species and its growth rate depend on only one variable, the total catch. In reality, of course, things are much more complicated than this. Thus the impact of a catch on the population of a species depends on:

- The age structure of the catch. Typically, if young specimens are over-represented in catches the effects on subsequent population levels will be more devastating. For this reason fisheries controls frequently specify minimum meshes of nets.
- Sex structure of catches. This is best illustrated with herd animals rather than fish. Deer herds contain large numbers of non-breeding males. Culling non-dominant (non-breeding) males will have a smaller impact on subsequent population levels than culls of females since all females breed and are serviced by dominant males. A policy of culling non-breeding males may

have long-run effects on the genetic strength of herds since competition for breeding is reduced but this is highly uncertain.

- Location and timing of catches. With pelagic fish, catches on the spawning grounds at times of breeding are likely to be especially devastating since in addition to the numbers taken the breeding cycle can be disrupted.

Thus a sustainable yield policy for a fishery or any other exploited species has in practice to address many more parameters than simply the total catch.

Summary

- Renewable resources are natural species of plants and animals that are economically exploited. Examples include fisheries and other animals hunted for food.
- The options for exploitation depend on the population dynamics of the species, which can be summarized in a sustainable yield curve.
- Populations are dynamically stable at levels above that which gives the maximum sustainable yield and dynamically unstable below that. A species is dynamically stable if in response to an increase in exploitation its population growth rate increases to compensate. Long-term survival of species is best safeguarded by keeping population levels in the region of dynamic stability.
- With secure property rights the probability of exploitation being kept at dynamically sustainable levels is greater:
  (a) the lower the discount rate used by the exploiter;
  (b) the steeper the increase in costs with falls in population levels;
  (c) the lower the elasticity of consumer demand for the catch.
- Open access exists where there are no secure property rights to the renewable resource. With open access exploiters take no account of future populations and yields since they have no ability to enforce any conservation strategy. Thus the probability of the population being driven to extinction is greater. Only factors (b) and (c) above serve to reduce the probability of over-exploitation.
- Protection of renewable resources requires a control authority to enforce property rights (restrict access) to the resource. If this is insufficient to ensure long-term survival of the resources, further controls may be necessary.
- Catch quotas or restrictions on fishing effort are the normal means of achieving this. These are open to the objection that they are not allocatively efficient. Tradable permits or tradable quotas have been proposed as a means of eliminating gains from trade and ensuring allocative efficiency.
- One strategy for sustainability of renewable resources is to farm the species concerned. This option is limited by the technology for controlling reproduction.
- A sustainable strategy for a renewable resource may entail much more than controlling the total catch to include the age and sex of specimens taken and the timing and location of the catches.

Chapter 15

Sustainability and sustainable development

In common with most national governments the UK Government is committed to sustainable development. It stated this commitment in the White Paper, *This Common Inheritance* (Department of the Environment, 1990), from which has followed a substantial number of official documents. A search through these sources for a clear definition of either sustainability or sustainable development, however, proves in vain and as noted in Chapter 1 there are a large number of definitions of the term.

But the search for precise definitions should not distract attention from a broad area of agreement. The issue of sustainability stems from a concern with the welfare of future generations and particularly with their access to the planet's life-support systems. This arises from a belief that current levels and forms of economic activity threaten those systems. Sustainability thus means constraining human economic activity so as to protect those life-support systems. Sustainable development is then economic development or growth (Box 15.1) subject to those constraints. What then are these life-support systems? The principal candidates, on which there is something approaching a consensus, are as follows:

- *Integrity of the atmosphere.* Current levels of pollution endanger the integrity of the atmosphere in two respects: the destruction of the stratospheric ozone layer, which provides protection against ultraviolet radiation; and the build-up of greenhouse gases resulting in global warming. Protection against ultraviolet radiation is necessary to sustain life in its current forms; life is only possible within a limited range, albeit ill-defined, of temperature and other climatic parameters. Global warming is mainly seen not as directly life-threatening, but rather, through its effects on sea levels, shifts in climatic zones and climatic turbulence, as making the planet less hospitable to life and thereby limiting life-styles, population levels and distributions. Losses of land in coastal zones and of rainfall in the major cereal-producing regions are the main perceived threats.
- *Biodiversity* plays a strategic role in human welfare since exploitation of wild plants and animals is a source of many major advances in a range of industries including pharmaceuticals, food technology and agriculture. An

Box 15.1 Sustainable development or growth

Economic development is the term used for the shift from traditional societies to capitalist ones and is used to describe the growth in Third World countries who, it is supposed, must accomplish this transition in raising the standards of living, measured as income per head, of their populations. Developed capitalist economies have experienced this transition and their increases in average income per head are described as economic growth. We thus have sustainable development because the principal problems that concern the parties to the sustainability debate centre on Third World economies. This is for a number of reasons. Third World countries possess the majority of the world's biological species and the largest surviving examples of natural habitats. Hence the form of their development is the critical issue in biodiversity conservation. Developed countries account for the bulk of resource use, e.g. consumption of fossil fuels, minerals and timber from natural tropical forests. They also are overwhelmingly the current cause of threats to the atmosphere from the emission of ozone-depleting chemicals and greenhouse gases. The Third World on the other hand is the major source of human population growth. Should its peoples achieve even half the consumption levels of the Western developed economies, the threats to the global atmosphere and the other life-support systems would became vastly worse. Hence the key to sustainability is to shift the development path of the Third World away from the high resource using, high polluting patterns of the West. The choice of the term 'sustainable development' could also be seen as a recognition of the fact that since the vast majority of future human beings will live in the Third World they will be the principal beneficiaries of sustainability policies. Some authorities prefer to use the term 'development' to imply that sustainability will require substantial changes (transitions) of lifestyles away from resource-intensive consumption within developed countries.

understanding of the complex interrelations between components of ecosystems, together with exploitation of genetic material, forms the basis of developments in biotechnology and offers progress on the control of agricultural pests, the control and treatment of disease, sustaining and advancing agricultural technology, etc. For these reasons protection of the genetic stock, avoiding extinctions of plants and animals and maintaining ecosystem diversity is also a requirement of sustainability.

- *Stocks of exhaustible resources.* These are discussed in Chapter 13, which demonstrates that sustainability requires improvements in the efficiency of use and extraction, and that the development of renewable and reproducible alternatives should leave successive generations facing no greater constraints on the availability of exhaustible resources than are experienced by the current generation.

- *Renewable resources* of current economic value, e.g. marine fish populations. Sustainability requires that exploitation be kept within dynamically sustainable yields.

Beyond these requirements of maintaining atmospheric integrity, genetic diversity and exhaustible and renewable resource availability, there are what might be described as primarily national desiderata and thus dependent on the particular circumstances of the country concerned. These include purely cultural objectives, such as the conservation of landscapes and historical and archaeological sites and structures. The justification for their inclusion rests on the argument that human welfare and cultural progress build on the past, and that an access to that past and an understanding of it requires the conservation of these features.

Cultural sustainability objectives, while differing in details between parts of the world, are probably universal. Other national sustainability constraints are dependent on the particular conditions of the country or region concerned. Thus a country subject to population pressure might define a sustainability objective of conserving its agricultural land. Similarly there might be a requirement for the conservation of water resources and particularly for maintaining the integrity of groundwater sources since pollution of such sources can constitute a long-term problem.

The life-support systems can only be specified in fairly broad terms and a commitment to sustainability thus requires a debate as to interpretation, to determine what precisely the sustainability requirements are. While the global requirements, subject to scientific uncertainties, are relatively clear in global terms, translation of them to the region or country is a matter for debate on how the burden is to be shared. Sustainability is about *inter-generational equity*, of how non-reproducible resources should be shared between current and future generations. Consistency requires therefore that the issue of *intra-generational equity* in meeting the burdens of sustainability be addressed. Intra-generational equity has two dimensions: between countries, and particularly between developed and developing nations; and, within a country, between its citizens. For a relatively wealthy country such as the UK intra-generational equity requires at a minimum that its own use of resources should be sustainable and that the costs of ensuring this should be fairly distributed among its citizens.

Sustainability and natural capital

The complexity of defining the objects of sustainability, what should be safeguarded and set aside for the benefit of future generations, is avoided by neoclassical economics. Its approach to sustainability rests on the concept of natural capital.

Natural capital comprises the capacity of the environment to provide goods and services to human populations. It thus includes the capacity of environmental media (air, water, land) to assimilate the waste products of human economic activity; the capacity of the atmosphere to sustain life, to keep it warm and shield it from harmful rays; the fertility of the soil, natural plants and animals with their potential to contribute to the maintenance and growth of human welfare, the supply of minerals and fossil fuels; everything indeed listed above as components of the life-support systems for future generations.

The production of goods and services requires, in the neoclassical approach, not simply the conventional factors of production (land, labour and reproducible capital) but natural capital as well. In the neoclassical world factors of production may be substituted for each other and techniques of production differ in the relative proportions of factors that they use. Thus one can substitute reproducible for natural capital in production and, since natural capital is heterogeneous, different types of natural capital for each other.

The argument then is that, because natural capital is frequently unpriced (thus we do not (yet) pay directly for the air we breath and the fisherwoman does not pay for the fish she takes out of the sea), society chooses techniques which utilize too much natural capital and too little reproducible capital and labour, which have market prices and hence have to be paid for. Indeed it leads us to consume too much and to invest too little. We observe the growing income per head and the growing stocks of reproducible capital and fail to notice the decline in natural capital. Our growth in welfare is less than we suppose. In the process of raising measured per capita income we are destroying irreplaceable but necessary natural assets. As many people do when they grow old, societies are living off their capital; the difference is that society does not recognize that it is doing it.

This diagnosis leads to two sustainability criteria:

- *Weak sustainability.* Natural capital should only be consumed in so far as it is compensated for by increases in the stocks of reproducible capital.
- *Strong sustainability.* The quantity of natural capital should be maintained constant; depletions of some sorts of natural capital should be compensated for by increases in other sorts.

Weak sustainability is satisfied if two conditions are met:

1. Environmental effects of private consumption and investment decisions are internalized by the use of Pigovian taxes.
2. All public investment projects achieve a positive present value when the environmental consequences are given their proper monetary value and included in the project appraisal.

This second requirement is because the value of the reproducible capital that the project creates is the present value of the income stream, net of the project costs that it generates. If the value of the environmental impacts is included in the costs and the project still yields a positive present value, then it has added more to reproducible capital than it has destroyed of natural capital. This relationship with weak sustainability is of course part of the reason why neoclassical economists have been so enthusiastic about placing monetary values on the environmental impacts of public investment projects.

Strong sustainability requires a different criterion, namely that the sum of environmental impacts of public investment projects is positive or at least non-negative. If the impacts are negative then what is called a shadow project is required to create positive environmental effects to offset the negatives of the project.

Box 15.2 Green social accounts

Conventional national income accounts present the annual flow of income for an economy. Gross domestic product (GDP) is the sum of the value added of all the enterprises located within the geographic boundary of the nation. Because this value added accrues as payments to the owners of factors of production within the economy it is also the sum of wages, profits, interest and rents received for activity within the national boundaries. Gross national product is a comparable aggregate but calculated for all the assets owned by the citizens of the nation. Because people can own assets abroad it differs from GDP. Net (domestic or national) product (NDP) is gross product less an estimate of the extent to which capital assets have depreciated (through wear and tear) in the course of producing the product. Capital consumption, as this is called, is conventionally confined to reproducible capital. The case for green accounts has two elements: (1) natural capital yields income that is not measured in GDP (free recreation etc.); (2) since natural capital does not have a market value and is not recognized in the conventions of social accounting its depreciation is not accounted for. On the former point there are a number of sources of income that are not recognized in social accounting conventions – unpaid domestic labour is one which provokes periodic controversy. The second argument is that when natural capital is a factor in the production of goods and services the income flow it yields is part of the GDP but the consumption of natural capital is not deducted when calculating NDP. Capital consumption in principle is measured as the difference between the value of the stock of capital assets at the end of the accounting period less the value of those same assets at the beginning of the period (there are complications if the price level changes over the accounting period, but we ignore these). As an example, if a natural forest is felled and the timber sold, the value of the timber will be part of the value added but no deduction for the decline in the timber stocks will be made in the accounts. When a commercial forest is felled there is identifiable capital consumption since the measured value of the timber stocks declines. Similarly if production of a good pollutes a river, the depreciation in water quality will not feature in the national accounts. If money is then spent by someone else in cleaning up the river that expenditure adds to GDP. Greening the accounts would require a beginning and end of year valuation of the natural capital stock. That is not feasible. What can in principle be identified is the component of GDP derived from protecting and replenishing natural capital, (pollution control, environmental protection, etc.) but in the absence of information about the aggregate state of the natural capital stock this tells us very little. The issue of green accounting is technically very complicated and the issues are not treated in this book.

The neoclassical doctrine has also created a momentum for green social accounts to supplement conventional national income accounts (Box 15.2).

The neoclassical approach to sustainable development depends on the meaning attaching to the concept of natural capital and the possibility of valuing it. It requires that the various components of natural capital, the functions of ecosystems and of the atmosphere, may be valued so that alternative configurations may, at least conceptually, be compared. As argued in Chapters 5 and 11 this is not possible. Natural capital is an aggregation of a wide range of disparate things, many of which cannot even conceptually be given any sort of monetary value. There is no correct answer to the question of whether a section of a trunk road is worth more than five sites of special scientific interest (SSSIs), the remains of an Iron Age field system, the sites of a Roman villa and a Civil War battlefield and an addition to the rate of emission of greenhouse gases. Society has to make such judgements and it is better that they be done consciously than by default. But to pretend that the problem can be solved by cost–benefit analysis arrogates to economics and its valuation techniques tasks and capacities it cannot satisfy.

Reproducible capital itself is heterogeneous, consisting of assets ranging from factories, shops, offices and warehouses to all kinds of plant and machinery, vehicles, standing crops and livestock, growing forests, power stations and computers. During the 1970s there was a long and, at times, acrimonious debate between neoclassical and non-neoclassical economists as to whether meaning could be given to the concept of aggregate reproducible capital. The outcome was a draw: neoclassical economists continued to believe in the notion since it was necessary for their explanations of how markets worked and their theories of economic growth; non-neoclassical economists continued to consider the notion meaningless.

The value of an existing item of reproducible capital is the present value of the income stream it will yield to the owner. As in the discussion of the investment decision of the private investor in Chapter 10, this will depend on two elements: the forecast profits that can be earned by the use of the asset and the rate of discount used for calculating the present value of those forecast profits. The present value will vary with the discount rate and as the discount rate falls the present value of long-lasting assets, i.e. those yielding profits in the more distant future, will rise relative to shorter-lived assets (see the discussion of Fig. 10.1). For a given set of capital assets an aggregate value can be calculated for each rate of discount, although whether the aggregate value will rise or fall as the discount rate rises will depend on the profiles of the anticipated profit streams. For any given discount rate the aggregate value will vary as owners alter their forecasts of future profit streams. Thus there is no unique value for the aggregate stock of reproducible capital independently of market conditions.

When we consider the notion of the aggregate stock of natural capital the conditions that permit some form of aggregation of reproducible capital do not exist. It is not possible to calculate the present value of the ozone layer or a redshank at any discount rate since forecasts of future profit (or income) streams from those assets (if indeed such they are) cannot be made. The 'measuring rod of money' cannot be used to aggregate items of natural capital. There is no answer to questions of the form: how many redshanks equal a ton of CO_2 discharged to the atmosphere or how many miles of road equal a site of a Roman villa? In the absence of any way of adding up quantities of disparate entities, the concept of aggregate natural capital is simply meaningless.

A standards approach to sustainability

Fortunately sustainability does not require the concept of natural capital. Once the commitment to sustainable development is made, sustainability requirements can be expressed as a set of constraints on decision-making. Thus if it is decided that the sustainability objective of safeguarding biodiversity is achieved by protecting designated wildlife sites (e.g. the set of SSSIs), then this is expressed by a requirement that investment projects avoid SSSIs or that they can only damage them if alternative and comparable habitats are created to replace them (if this is technically possible: in practice with current technology it usually is not). Similarly if safeguarding the integrity of the global atmosphere requires limiting the emissions of greenhouse gases to pre-specified quantities, then either investment projects have to keep within these constraints or they must be accompanied by action to save greenhouse gas emissions elsewhere in other sections of the economy. These constraints on economic activity take the form of environmental standards and give us the standards approach to sustainability.

Environmental standards can take a number of forms depending on the nature of the sustainability commitment and the constraints that it implies. They might be classified as follows.

Discharge standards

These are familiar from the discussion of the problem of pollution control and relate to the emission, to any receiving medium, of waste products from economic activity. As shown in that discussion, the specification of standards is central to pollution policy and specifying a standard does not imply that any particular instrument should be used to obtain it. A wide range of economic and command and control instruments is feasible and the choice of control mechanism will depend on the nature of the discharge.

Stock standards

Where the requirements of sustainability mean that certain specific sites be safeguarded, e.g. valuable wildlife sites necessary to meet the objective of safeguarding genetic diversity or critical cultural assets (Stonehenge), stock standards will apply. A stock standard applies to irreplaceable assets deemed to be vital to meeting the commitment of sustainability. With a stock standard certain types of economic activity that will damage or destroy the protected asset are prohibited on that site. Examples might be prohibition on routing new roads through such sites or building factories or power stations on them. With stock standards the principal instrument of protection is the planning system, although where the protection of the site requires positive action on the part of the owner other instruments may be necessary. An example discussed in the next chapter concerns the protection of semi-natural habitats in the UK, where the wildlife value is dependent on the continuation of specific agricultural practices.

Flow standards

There are a lot of cases where sustainability entails maintaining a given quantity of a type of asset but not the existing specific examples of it. In these circumstances flow standards will apply. A requirement for a flow standard is therefore that the technology exists for creating alternatives to the existing examples. Since flow standards have a wide range of application some examples will illustrate.

1. For types of habitat where it is possible to create alternatives, or to transfer the habitat to a new site, a flow standard might be used to meet those aspects of sustaining biodiversity. A flow standard would require that an existing habitat should only be damaged or destroyed if investment in habitat creation or relocation were made to offset the loss. This investment might be in restoring degraded examples of the habitat type, in transferring valuable biota to new sites or in creating habitat *de novo*.
2. Where the total area of habitat of a specific type is greater than is deemed necessary for safeguarding it for future generations and only examples are currently protected, a flow standard might require simply the designation and protection of an alternative. Scarce types, where alternatives do not exist, require either relocation or creation of an alternative. If that is not possible then they need to be protected by a stock standard.
3. A similar approach could be made to maintaining landscape quality in all but the highest quality of landscapes, which are unique and irreplaceable. Damage to landscape from some types of economic activity can be offset by investment in restoring and enhancing degraded landscape elsewhere (see Bowers and Hopkinson, 1994).
4. Flow standards might equally be applied to retaining examples of buildings and archaeological sites for future generations. Again the standard would require that protected examples could only be destroyed or damaged if other examples were protected to replace them.
5. A flow standard might also apply to exhaustible resources. An example might be the depletion of fossil fuels. This should be offset by investment in the technology of renewable sources of power and in techniques to increase the effective capacity or efficiency of remaining stocks (energy conservation, etc.). A flow standard would require that the investment be functionally related to the rate of depletion of fossil fuels and not simply market determined.
6. Finally flow standards could be used for meeting targets for the control of global pollutants where the location of the discharge is irrelevant. Thus a project that caused an increase in CO_2 emissions could be offset by action to save a comparable quantity of emissions elsewhere in the economy.

A flow standard imposes an obligation on the instigator of the unsustainable activity to make the investment necessary to neutralize it. If the activity requires planning consent then the standard may be maintained by making the investment a condition of consent. But the most general instrument for flow standards is the levying of a *sustainability tax*. A sustainability tax has two essential features:

1. It is based on the cost of maintaining the standard, on the costs of the work or action needed to neutralize the effects of the damaging activity. It is not related to the value generated by the activity causing the damage nor to any monetary valuation of the damage.
2. The revenue from the tax is *hypothecated*, i.e. it is spent on the purposes for which it is levied. Governments usually resist hypothecation because it limits their freedom on spending decisions and on grounds of economic efficiency: that the hypothecated purpose might not be the most effective use of the funds generated. But the purpose of the tax is to maintain the sustainability standard and if the revenue is not spent on that purpose the standard is not being met.

An example of the use of a sustainability tax to meet a flow standard for landscape quality is described in Box 15.3. The same idea might be applied to the

Box 15.3 A sustainability tax for landscape quality

The proposition rests on the ability of the landscape architectural profession to grade landscapes and to appraise potential developments in terms of their impact on landscape quality. A study would be made of the typical means of landscape restoration for different types of development project and of the normal costs of this restoration. This would yield a scale of landscaping expenditure designed to harmonize structures of various kinds into landscapes of various types and qualities. The scale would be progressive since, for low landscape qualities, many forms of investment would not degrade, while for the highest qualities any intrusion would lead to a loss of quality. Thus for road investment the types of restoration works would grade from expenditure on tree planting and rules about slopes of cuttings to rules concerning colour and materials for bridges, up to the requirement for tunnelling to remove the structure entirely from the highest grades of landscape. This standard scale would be used as the basis for a tax to be levied on developments that damaged landscape. The liability for tax would depend on an assessment of the deterioration of landscape quality that would result from the project. The proceeds of the tax would accrue to a statutory landscape body responsible for seeking and carrying out landscape improvement projects or alternatively to be deposited in dedicated landscape accounts of local government bodies. Because the landscape standard is a flow standard there is no requirement that the landscape improvements should occur in the areas that were being damaged. Indeed there was a case for concentrating the investment on the most degraded landscapes. (Bowers and Hopkinson, 1994).

impact of projects on the emission of greenhouse gases with the proceeds of the tax devoted to projects to reduce greenhouse gas emissions. Other possibilities might be charges for wildlife damage or for damage to archaeological sites.

Costs of sustainability

With the standards approach to sustainability there is no need to value the environmental effects of economic activity in monetary terms and no need to invoke the notion of natural capital. This does not, of course, mean that sustainability does not have a cost. Since economic activity is constrained by the standards, the cost is the economic benefits forgone and the additional costs incurred as a result of meeting the requirements of sustainability. For the three types of standards the costs are as follows:

- *Discharge standards*: The output forgone and the pollution equipment needed to meet the standards plus the costs of monitoring and enforcement.
- *Stock standards*: very few economic developments can only take place at one location so a stock standard will normally cause the relocation rather than abandonment of projects. Alternative sites may involve developers in higher costs and be less profitable. Often, however, site choice depends on the accident of ownership rather than arguments about economic efficiency. In most industries, transport and other spatial costs are only a small fraction of the costs of production. In other cases, costs are minimized by recognizing the constraints at the appropriate stage in project planning. This is particularly the case with road construction, where the practice of appraisal in sections constrains route choice, resulting in 'unavoidable' destruction of wildlife and archaeological sites This was the situation with the Winchester Bypass, where the choice of line was reduced virtually to zero because the road had been built on both sides of the short stretch that involved the valuable environmental asset of Twyford Down. The effective choice was then reduced to either an expensive tunnel or a destructive cutting. Had the planners treated Twyford Down as protected at the strategic route-planning stage, the opportunity cost of protecting it would have been far lower. Hence the opportunity costs of stock standards are likely to be small.
- *Flow standards*: if sustainability taxes are used to enforce flow standards and they have been properly calculated, the tax revenues measure the costs of the standards. Where planning obligations are used, the cost of the standard is the costs incurred as a result of those obligations.

In principle a commitment to sustainability entails a sacrifice for current generations in the sense of forgoing current opportunities of economic development for the benefit of future generations. But if sustainable development requires the correction of existing market failures and an improvement in the efficiency of resource allocation, then there will be benefits to current members of society to offset the costs. Studies have identified specific cases in developing countries where greater care of the environment would improve income (a summary of some of this work is Pearce *et al.*, 1990) and the work of Repetto has suggested that inefficient policies of subsidization of pesticides and irrigation are widespread and have resulted in much unnecessary environmental destruction and loss of income (Repetto, 1985, 1986). The classic study of forestry policies by the same author identified multiple failures in the licensing and rent-extraction systems that led to excessive felling and degradation in most major forests including those of North America and a failure by the countries concerned to capture the value of their timber resources (Repetto, 1988). Similarly the Common Agricultural Policy (CAP) of the EU has resulted in food that costs more to produce than its market value. The community is made worse off by these policies even though farmers benefit. The destruction of the farmed environment has been incidental to the CAP. If agricultural production is curtailed in the interests of maintaining genetic diversity, landscape or some other sustainability objective, the costs to the community will be negative. Chapter 12 showed how the distortion of cost–benefit techniques resulted in many land drainage schemes that were not only environmentally damaging but did not meet the investment criteria. Similarly the techniques of trunk road appraisal led to environmental damage for what are frequently trivial time savings for motorists. In both cases curtailment for purposes of sustainability would have negative costs since the funds devoted to the investment would have been better spent in other ways.

On balance a commitment to sustainability seems likely to entail positive costs on current populations for the benefit of future generations. Sustainable development in the context of a developed country, such as the UK, means that economic growth is less than it would be otherwise. It is not currently clear what that level of costs will be.

Sustainability and environmental quality: is sustainability simply being 'green'?

The notion of sustainable development arose out of concern that economic activity was leading to excessive environmental degradation and many policies for environmental protection pre-date commitments to sustainable development. Sustainability policies seek to conserve the environment for the benefit of future generations. In the absence of sustainability, environmental policies seek to provide environmental 'goods' for the current generation. It is clear that the demand for environmental goods is *income elastic*, i.e. as people get wealthier they spend a higher proportion of their income and devote more of their leisure time to activities that involve environmental goods. Even without the Earth Summit, therefore, pressures for environmental protection would be high in Western countries, although not of course in the Third World where poverty dictates different priorities. The question that arises, therefore, is whether, for a developed country, a commitment to sustainability adds to environmental quality. Does it result in a better environment than a situation where there is a strong preference for environmental goods but no commitment to future generations?

There may be aspects of the environment that have no relevance to life-support systems but which have high current consumption value. Examples would include what I have termed cultural aspects of sustainability: landscapes and historic sites that are fully documented and wildlife and recreational sites of purely local value. However there are a number of reasons for supposing that a commitment to sustainability will lead to an increased rate of environmental protection over the alternative of enhanced green-ness:

1. Some forms of pollution will have little short-term impact but may have substantial negative long-term impact. This is the case with greenhouse gases, which may even yield short-term benefits (long hot summers in which to enjoy environmental goods) and may be the case with some forms of water-borne pollutants.
2. In some cases the current situation, while acceptable, is unsustainable so that sustainability dictates environmental improvements above what the public would otherwise demand. This is probably the case with biodiversity where wildlife conservation requires improvements in the wider country-side and not just the safeguarding of a few high-profile wildlife reserves.
3. Sustainability directs attention to aspects of environmental diversity that have little or no current consumption value. Thus biodiversity leads to concern with the needs of obscure and 'invisible' biota that may be endangered and not simply to 'high-profile' plants and animals for which there is a current public demand.
4. Sustainability is global and not simply regional or national. Beggar-my-neighbour policies, such as exporting toxic waste to Third World countries or acid rain to Scandinavia, or importing products derived from degrading environments in the Third World, are unsustainable but may serve to safeguard current consumption and environment in the West. This is not of course to say that sustainable development has eliminated these practices.

This leads us to reconsider the costs of sustainability. A distinction has to be drawn between the costs of meeting sustainability constraints *per se* and the costs of eliminating environmental degradation or making environmental improvements deemed necessary by current concerns unconnected with sustainability. Thus while it is clear that some sections of industry are investing substantial sums in cleaning up their act (ICI is reported to have spent over £200 million in 2 years on Teesside) this would probably have been necessary in any event because of public concern over the health implications of its activities.

Sustainability and discounting

There is a debate as to whether a society committed to sustainable development should discount at all, since discounting means giving lower weight to costs and benefits incurred in the future (and the further in the future they occur, the lower the weight given to them).

It is important to note at the beginning of this section that there is nothing that can be done about discounting by private individuals and companies. The discussion therefore has to be whether discounting should be practised in public decision-making.

Individuals discount for a number of reasons including the inescapable probability of death and hence uncertainty that the benefits/costs will be experienced. But society should assume that it will not die and discounting thus gives lower weight to the needs of future generations compared with current ones. However the typical investment project entails costs now that are borne by the current population and future benefits which accrue to future generations. Thus, if discounting is not practised, future generations receive benefits but do not incur the costs, while current generations incur costs but do not receive the benefits. It is therefore argued that inter-generational equity requires discounting. On the other hand, where projects impose only costs on future generations and yield benefits for current generations the case for discounting is weak. Nuclear power, with future problems of decommissioning and storage of nuclear waste, is an investment that has this property and special procedures may be required to deal with nuclear investments. Most public investment is not of this form.

The other argument for discounting is that society has limited resources at any time and a public investment has opportunity costs in the form of private consumption/investment forgone. A zero discount rate for the public sector would greatly increase the amount of public investment taking place (and hence the amount of taxation to pay for it) at the expense of other investment and consumption. Thus discounting is a device for rationing funds between the public and private sectors. A zero discount rate would lead to a much higher volume of current public investment and could easily therefore increase rather than reduce the rate of environmental degradation.

The standards approach to sustainability does not require any changes in the discount rate since the interests of future generations are protected by constraints on decision-making rather than changes in the values and weights given to the costs and benefits.

Local action and sustainability

The Brundtland Report and the Earth Summit placed stress on the importance of action by local communities in achieving sustainable development. These sources view sustainable development as a bottom-up rather than top-down process. The thinking underlying this view is developed further in Chapter 16 but some preliminary comments fit into this section.

The previous chapters developed what could be called the coercive model of environmental control. In this view a central authority specifies environmental objectives and seeks to achieve them by the use of instruments, either command and control or economic instruments. In the context of sustainability the coercive approach presents two broad problems.

First, since the objectives sought are presumed to be against the private interests of the parties who are being controlled, the policies carry what, in the context of sustainability, may be an unacceptable risk of failure. Second, where environmental problems are non-point, such as the cases of household waste creation and urban atmospheric pollution from short-distance car journeys considered in previous chapters and particularly where the environmental problems stem from the actions of individuals and households, top-down control may not be cost-effective. Voluntary action by individuals, which they will not undertake in a coercive system, may be effective.

As an example we can consider energy conservation. Modifications to household behaviour, such as switching off lights, installing energy-saving light bulbs and walking rather than driving to the local shop, may in aggregate be significant although each instance yields trivial energy savings. A control authority might try to influence energy savings by, for example, imposing an energy tax, but if each behavioural act by the household saves a trivial amount of energy then it saves a trivial amount of tax and households may not respond to the instrument. What we have here is a form of transactions cost between the control authority and the households created by the need for a large number of actions by the households if the target is to be achieved.

The alternative approach is to 'bond' the households to the objectives so that they carry out the action voluntarily without the need for coercive instruments. Local action seeks to elicit the support of local communities for sustainable development. It does this by local participation in defining the objectives of sustainability and in determining the means of meeting the objectives. The idea is that through participation local communities, and the households that comprise them, acquire property rights in sustainable development. The economic benefits of local action are as follows:

- the chances of individuals adopting strategies that are in conflict with the objectives are reduced;
- the ability to identify problems and to solve them is increased since many more eyes and ears are brought to bear;
- problems of intra-generational equity are reduced since participation increases the willingness to make sacrifices and self-imposed behavioural changes are accepted.

Seen from the centre, local action has also a cost: that the objectives of sustainability will get redefined in the participative process and the priorities altered from those that the experts at the centre would choose (for examples of this process of modification see the symposium on Local Agenda 21, *Ecos*, 1996).

Sustainability and ecocentrism

Economics by its very nature has a human focus; it is concerned with the ways in which human societies can utilize the resources at their disposal to maximize their welfare. Environmental economics does not differ in essence from this perspective; it gives economics a different emphasis, placing stress on the importance of the environment to human welfare and on how neglect of the environment can undermine that welfare. Environmental economics therefore naturally adopts anthropocentric definitions and perceptions of sustainable development. There are schools of thought in the sustainability debate that question this perspective and start from an ethical position that other animals have equal rights to existence and presumably therefore the means to existence as humans. It is not my intention to examine the ethics of these positions, rather I consider their implications for the notion of sustainability and sustainable development.

One must be clear whether ecocentric doctrine applies to individual animals or to species (or distinct populations of species). It is necessary also to clarify what is meant by animals. Thus if what is being asserted is the rights of individual animals and 'animal' is interpreted to mean a member of the animal kingdom (as opposed to the plant kingdom and microscopic organisms: bacteria, viruses, etc.) then any human action probably inadvertently destroys animals and their habitats. The realistic interpretation is probably that ecocentrism is intended to assert the principle that economic activity should be so constrained as not to damage or destroy wildlife and its habitats where wildlife is anthropomorphically defined (the local nature reserve, not the insects in the polluted wayside puddle).

If that is a correct interpretation, the neoclassical perspective is clearly in conflict with it since, under weak sustainability, it allows that habitats may be destroyed and species even rendered extinct if, in a cost–benefit analysis, when the wildlife habitats are valued by a contingent valuation survey or in some other way, the measured monetary benefits outweigh the costs. Under strong sustainability the same might apply, except that the losses would require some compensating investment in natural capital, which from the viewpoint of the affected biota, unfamiliar with the concept of aggregate natural capital and whose habitat is lost, is probably not much consolation.

The standards approach, on the other hand, is not so clearly in conflict with ecocentrism. Depending on what is encompassed by environmental standards, the approach requires that economic activity be so constrained and structured that a representative set of wildlife habitats is safeguarded or any damage is compensated for by habitat creation or designation of alternatives. This is probably a realistic compromise with ecocentrism. Even with zero economic growth, artificial structures will wear out and need to be replaced, and given the ubiquity of wildlife habitats some examples will be lost just in the replacement process. It is incumbent on societies to so structure the sustainability constraints that nothing in their judgement, informed by the relevant science, is lost or damaged.

The next chapters examine the question of what difference a commitment to sustainability makes to environmental policy choices by looking at two issues: biodiversity and global atmospheric pollution.

Summary

- Sustainable development is development or growth that satisfies the requirements of sustainability.
- While there is argument about the precise definition of sustainability there is consensus that it entails passing essential systems intact to future generations.
- Sustainability is concerned with inter-generational equity but implementation of intra-generational equity requires the distribution of any sacrifices required for sustainability.
- The basic life-support systems are the integrity of the global atmosphere, biodiversity, and stocks of exhaustible and renewable resources. They may also include water resources and stocks of good agricultural land. In addition most nations will define cultural sustainability

objectives such as heritage landscapes and historical sites. Life-support systems can only be defined in broad terms and there is debate as to the precise requirements of sustainability.

- Neoclassical economists incorporate all of these life-support systems into a concept of natural capital and define sustainability in terms of this concept. This leads to two notions of sustainability: weak sustainability and strong sustainability.

- Weak sustainability is maintained if the stock of natural and artificial capital is constant or increasing through time. Strong sustainability requires that the stock of natural capital be kept constant or allowed to increase.

- Weak sustainability is satisfied if all environmental effects of private decisions are internalized through a set of Pigovian taxes and public investment has to satisfy a cost–benefit test when environmental effects are incorporated and given monetary valuation.

- Strong sustainability requires that any losses of natural capital in public investment projects be compensated for by shadow projects that create natural capital of equal value.

- The concepts of strong and weak sustainability rest on the notion of natural capital. This in turn requires that it is possible to place monetary values on different environmental impacts of investment so that the net impact may be compared with the artificial capital created (weak sustainability) or the compensatory investment determined (strong sustainability).

- There are many environmental effects that may not be given meaningful monetary values. Hence the notions of weak and strong sustainability, which depend on monetary valuation, are not operational.

- The alternative approach to sustainability is to incorporate sustainability requirements as environmental standards that serve as constraints on economic decisions.

- Environmental standards can be of three basic types: discharge standards, stock standards and flow standards. Discharge standards control the discharge of waste substances to the environment. Stock standards protect irreplaceable assets and require that economic decisions are designed so as to avoid damaging them. Flow standards require compensating environmental improvement to offset damage done but, unlike strong sustainability, the compensation is physical not monetary. Destruction of a habitat requires the creation of one of the same type or class.

- Nonetheless flow standards can be implemented by levying sustainability taxes on development. A sustainability tax is calculated on the basis of the cost of maintaining the standard, i.e. correcting the environmental damage, and *not* on the basis of monetary valuation (willingness to pay) of the environmental damage.

- Sustainability taxes are hypothecated. The revenue is spent on maintaining the standard by correcting the damage that the development has caused.

- The standards approach to sustainability does not require monetary valuation of the environment. Nonetheless it entails costs that are incorporated in the costs of the projects to which standards apply.

- A number of writers have argued that sustainable development confers benefits on current generations as well as future generations. It does this by correcting policies and practices leading to unnecessary environmental damage. But while the costs are difficult to determine, a commitment to sustainable development probably does impose costs on current generations. Hence the issue of intra-generational equity is important.

- Even without a commitment to sustainability there would be public demand for environmental improvement. Sustainability probably increases the amount of environmental improvement that is required.

- Some writers have argued that sustainability requires the abandonment of discounting which, by its nature, discriminates against future generations.

- In their private economic activity people discount the future. A zero discount rate could only be operated for public investment. It would have the effect of increasing the volume of public investment relative to private investment and consumption. It is not clear that this would benefit the environment. Neither the neoclassical nor the standards approach to sustainability require a zero public sector discount rate.

- The Earth Summit saw the need for local community action to achieve sustainable development. Local action increases the acceptability and attainability of sustainable development but the sustainability requirements may be redefined in the process.

- Economic approaches are human centred. There are alternative ethical viewpoints termed ecocentrism. The standards approach constitutes a working compromise with these viewpoints.

Chapter 16

Instruments for biodiversity policy

Much of the discussion of instruments for environmental policy is applicable to a context where there is a commitment to sustainable development. There are, however, two areas where different considerations apply: biodiversity and problems of global pollution. These are the subject of this and the following chapter.

Biodiversity and biodiversity policy

There is an extensive and lively debate among biological scientists about the meaning of biodiversity and how it is to be measured (an excellent summary of the issues is the preface in Hawksworth, 1994, pp. 5–12). There are three forms or levels of diversity that are considered in this debate, ecosystem diversity, species diversity and genetic diversity, and the choice between them affects not only issues of measurement but also the specification of policy objectives and the choice of instruments. Unfortunately there is not space within this work to consider these issues and I must perforce proceed on some simplifying assumptions. I identify biodiversity policy broadly with species diversity but modified to incorporate recognized subspecies, races and distinct and separate populations of species of plants and animals. The objective of biodiversity policy is therefore to ensure the survival of the maximum number of species. A central and necessary (but not sufficient) condition for achieving this objective is taken to be habitat conservation, maintaining sufficient examples of viable habitats of each type to give an acceptable probability of long-term survival of the species they contain.

A biodiversity policy so defined will require the identification of key habitats and methods for their protection. This will obviously include protection against economic development, e.g. drainage of wetland habitats, ploughing up and afforestation of grasslands, moorlands and peatlands, felling of natural and semi-natural woodlands etc., and the construction of roads and buildings on protected areas. But it will also require positive management. Thus in Australia the key to maintaining indigenous wildlife is seen to be the control of alien species (rabbits, foxes and goats). In Western Europe many of the semi-natural habitats are artefacts of particular farming systems and their survival requires the continuation of traditional husbandry. Success in meeting biodiversity objectives requires also the development of technology for habitat restoration.

Biodiversity policy thus is likely to involve a range of instruments and much of what has been said about instrument choice in this book is applicable to instrument choice for biodiversity policy. But biodiversity introduces new considerations. It does this because of the problem of irreversibility. Few habitats, once lost, can be re-created in anything like their original richness, and plants and animals once driven to extinction cannot be re-created at all. Thus in the choice of instruments a premium is put on the avoidance of failure. This was not a consideration in our previous discussion and we noted, for example, that pollution standards are often exceeded despite the presence of sanctions or economic instruments designed to keep discharges within the chosen limits. Equally, ancient woodlands are felled, wetlands drained and protected areas built on.

Risk of failure and safety margins

The normal approach to reducing the risk of failure is to incorporate a safety margin in the setting of the levels of control instruments, and the adoption of a biodiversity strategy should increase the desired safety margin. The simplest illustration is the standard command and control model of discharges of a single pollutant into a receiving medium from a series of known point sources, which is considered in Chapter 6. In the discussion of command and control in that chapter the control authority set a standard S_c on the polluter, designed to achieve its overall pollution control standard. In the present instance it deducts a safety margin d from S_c and sets the polluter the standard S_c-d. The more averse the authority is to breaches of the consent standard S_c, the larger the value of d and the lower the consented level of pollution.

If control was by tradable permits the safety standard would be achieved by issuing permits for pollution of S_c-d. With a pollution tax the tax rate would be set to achieve S_c-d rather than S_c.

The use of a safety margin has a social cost for two reasons:

1. the control authority has to monitor for lower levels of discharges, which will raise the costs of monitoring;
2. the polluter is faced with greater levels of abatement, meaning either greater expenditure on control equipment or production at a lower level of output.

Some commentators have suggested that if, instead of lowering the standard, the penalty for exceeding the standard is increased, then the additional safety may be achieved without resource costs. This might be termed a financial as opposed to a physical safety margin. For command and control this idea can be illustrated using the presentation of the penalty system as an equivalent tax as developed in Chapter 6. The effective tax for discharges exceeding the consented level S_c is:

$$c(F) = p(D)p(P)p(C)e(F)$$

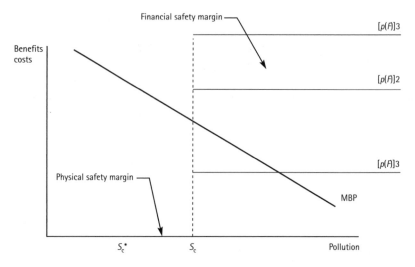

Fig. 16.1 Command and control as an equivalent tax.

and for given values of the probabilities of detection, $p(D)$, prosecution $p(P)$ and conviction, $p(C)$, the safety margin is achieved by raising the tariff of fines and thereby increasing $e(F)$. Figure 16.1, adapted from Fig. 6.2, illustrates the idea. $p(F)2$ is the preferred penalty function in the absence of the problem of irreversibility, but when irreversibility is recognized as a problem the higher penalty function $p(F)3$ is preferred. $p(F)3–p(F)2$ is thus the financial safety margin. It involves no resource costs since the polluter's output and the monitored pollution level is the same for $p(F)3$ as it is for $p(F)2$.

Since tradable permits, equally with command and control, require the backing of the law, financial safety margins could be used for them as well. What would be required is simply an increase in the penalties for polluting when not possessing the requisite permits. However, financial safety margins could not work with pollution taxes since the polluter may discharge at any rate she wishes provided that she pays the requisite tax. Tax evasion is a problem that has to be guarded against and which may be obviated by increasing legal penalties, but it is not the primary problem.

The extent to which safety can be achieved through adjustment of the penalties however is in practice likely to be severely constrained by jurisprudential factors. The penalty must be perceived by society as reasonable given the nature of the offence, otherwise either the probability of conviction $p(C)$ will fall to compensate (if the penalty for stealing a pocket handkerchief is deportation to Van Diemen's Land then jurors will not convict) or the penalties imposed will be what is seen as just and not what is allowed for by statute. Thus ensuring that irreversible environmental changes are avoided by levying enormous fines or threats of imprisonment (or execution!) of offenders is not a realistic option and physical safety margins, with consequent real resource costs, are necessary. Physical safety margins are necessary in any event to guard against 'no fault' or accidental infringement of standards.

The 'first mover' problem

While I have noted that breaches of control levels occur in practice I have not discussed why this is so. One obvious explanation is that the control authority got it wrong; it misunderstood the economic conditions under which those it was trying to control operated, and hence underestimated the incentives on them to ignore the controls. But there is another reason, which gets to the core of the problem of protecting against irreversibility. Instruments for maintaining biodiversity have to operate into an uncertain future. An instrument designed to deliver the objectives of the policy under current conditions may require revision when those conditions change. The revision required may be no more than marginal adjustments to control settings (tax rates, subsidy levels, tradable quotas or whatever) but it could be more fundamental, entailing a change in the set of control instruments in use. In designing a system for implementing a biodiversity strategy a responsible authority is operating under complete, or at least partial, uncertainty: the range of possible factors that could require changes to the system cannot be specified and hence the probabilities attaching to those factors are unknown.

Uncertainty thus attaches to factors that could, in various ways, offer new economic opportunities either to those at whom the control instruments are directed or even possibly to others who are not affected under current circumstances, and the exploitation of those economic opportunities will put biodiversity at risk. If the control authority could be certain that it would detect these new factors sufficiently far in advance to adjust its instruments to neutralize the threat, this uncertainty would not matter. But of course it cannot have this guarantee. The possibility that those who can benefit from the new circumstances will be the first to perceive the opportunities and exploit them before the control authority can act must be faced.

If knowledge of economic opportunities is not simultaneously acquired by all participants in a market then it is said to be *impacted*. Those who acquire the knowledge first are in a situation of temporary monopoly from which they may profit. The Austrian School of economists sees all monopolies as temporary in this sense and the exploitation of impacted information as the principal source of economic progress. The individual that acquires impacted knowledge is said to possess a *first mover advantage*. He or she has the opportunity to act on that knowledge, before the control authority catches up and thus to profit at the expense of biodiversity. If others perceive these opportunities before they are perceived by the control authority then they may damage biodiversity before the authority can act. This is the first mover problem. The protection of biodiversity, because of irreversibility, requires protection against first mover advantage.

The control authority has two broad strategies to cope with the first mover problem:

1. Invest in detection, monitoring and information gathering so as to reduce the risk of not being the first mover. There are two dimensions along which the authority might act, detection probability and response time, given detection.
2. Design control instruments that reduce or eliminate the first mover problem.

The detection probability is increased by extending the range of data collected to include the variables believed to determine the decisions of the people or organizations subject to control under the biodiversity policy, and utilizing these data to model their decision processes. Response times will vary between instruments. Some may require legal processes to bring about change, while in other cases fiat may be possible. In general informal instruments are likely to prove the most flexible but they may be less reliable in other respects.

What is required of an instrument is that when an individual whose activity is being controlled to conserve biodiversity, whom I term the agent, perceives a profit opportunity, she does not become a first mover, i.e. her reaction is to inform the control authority and seek to re-negotiate with it, before, or instead of, exploiting that opportunity.

Clearly not all possible instruments have this property. Thus where a firm is constrained from polluting a lake by a pollution tax, if the value of the activity that leads to the pollution increases, so that it becomes profitable to pollute the lake and pay the tax, it is unlikely to ask the pollution control authority to increase the tax. Equally if a landowner is prevented from cutting his forest by the knowledge that to do so renders him liable to prosecution with, if convicted, a fine greater than the value of the timber, a rise in the price of timber may cause him to take the risk but it will not lead him to campaign for higher fines. If, instead, protection is by a system of tradable permits, first mover advantage might lead the firm to bid up the permit price to acquire extra permits but if the incentive is strong enough the alternative is to exceed permitted levels and pay the fines incurred. All of these instruments are coercive and impose penalties on agents for going against the objectives of the control authority. Such instruments are called negative instruments. Some conclusions can now be reached. *Negative control instruments are not first mover proof.* Since negative instruments are predicated by the polluter-pays principle, while positive instruments (subsidies, property right allocations) are in conflict with it, *the application of the polluter-pays principle in the design of instruments places environmental assets at risk from first mover problems.* The third conclusion follows: *positive control instruments, those which offer some benefit in return for protecting the environment, afford at least some protection against first mover advantage.*

An instrument that offered exact compensation for losses incurred as a result of environmental conservation might afford some protection against first mover advantage but could well not do so. An example would be the system for the protection of agricultural sites of special scientific interest (SSSIs) in the UK, where farmers receive payment equal to the profits they would earn from exploiting and thereby destroying the SSSI (by ploughing it up, draining it, applying artificial fertilizers, etc.). Payment is based on a contract between the wildlife protection agency and the farmer. If the circumstances change so that he is making losses from sticking by the agreement he might seek to re-negotiate it. On the other hand this could be a time-consuming operation and the new profitable opportunity might be of only limited duration. Since in the long run he is neither better nor worse off by the agreement, he has an incentive to exploit his first mover advantage. The incentive to exploit first mover advantage is reduced, therefore, if the farmer receives additional payment in recognition of his contribution to the environment, since then, if the profitable opportunity is temporary, he may be worse off in the long run.

The best protection against first mover advantage is given by instruments that confer some property rights in biodiversity on the agent in a contractual setting. Biodiversity is a non-excludable good and everybody therefore has an inalienable share in it. A property right allotment thus cannot be made to biodiversity *per se*. However, the biodiversity strategy is composed of separate policies for individual habitats and sites and the control takes place at that level. *A property right allotment to the habitat that is subject to control has the externality of tying the agent to the biodiversity policy of which it is a part, thereby raising her valuation of biodiversity and reducing the risk of first mover exploitation.* The property right might involve the individual in charging for entry to a site or offering guided wildlife tours. Alternatively it may permit exploitation of wildlife resources in ways, and at levels, that do not endanger their survival.

The typical property rights instrument contains elements of contract. Residual rights are retained by the control authority; the agents retain limited property rights in the exploitation of the resource and are contractually bound to the control authority and, where these rights are held in common, as will often be the case, to each other. Property rights instruments of this sort have been developed for the control of fishing rights and for agricultural and forestry activities in tribal lands.

In summary, therefore, protection of biodiversity requires a rethinking of the traditional arguments for instrument choice and argues for controlled exploitation rather than absolute protection.

Biodiversity and moral hazard

Aversion to the risk of irreversibility means that instruments for biodiversity control must also carry some protection against what is known as moral hazard. Moral hazard arises whenever the outcomes of the actions of an agent are dependent in part on exogenous factors, 'states of nature', which the control authority cannot wholly observe. A practical (Australian) example might be where protection of indigenous species of animals depends on control of alien species, but population levels of both indigenous species and aliens are in part determined also by uncontrollable climatic and other factors, which may not be fully understood. The control authority could require the land-holder to control aliens, but it has difficulty in monitoring his activities except by their outcomes in terms of the survival of the indigenous species and the numbers of aliens present. The land-holder might choose to neglect his duties and blame the adverse results (declines in indigenous species, expansion of numbers of aliens) on the other imperfectly understood exogenous factors (thus he might argue that drought has caused the indigenous species to suffer and the aliens to prosper). This is the moral hazard problem. Enforcement of the duty through the courts is vitiated by the difficulty of proving cause. Economic instruments may also suffer from the same problem.

The conventional solution to moral hazard is for the land-holder to receive payment dependent on the survival of the target indigenous species. This is again a property rights instrument that ties the land-holder to the biodiversity objective. It also involves him in sharing of the 'state of nature' risk with the control authority. If the indigenous species decline despite the land-holder's efforts to control the aliens, then he suffers a loss. For this reason property rights systems will never entirely eliminate the risk of moral hazard, since the agent cannot capture all of the benefits of his contribution to biodiversity. Equally, and for the same reason, constructed property rights systems cannot entirely eliminate the risk of first mover exploitation. Indeed neither problem can ever be completely eliminated and biodiversity cannot be guaranteed against human action any more than against natural disasters, but the risks are reduced by recognition of the problems in the design of instruments.

Local action and biodiversity policy

At the end of the last chapter I argued that the objectives of sustainability are furthered by a programme of local action that gives property rights to individuals as members of a local community in sustainable development. In this chapter I have argued that protection against first mover problems and moral hazard in biodiversity policy is achieved by instruments that give limited property rights to the agents. In the former case, the objective is to 'bond' people to sustainable development as a whole; in this latter case, property right allotments serve to 'bond' agents to a specific sustainability objective, namely the conservation of biodiversity. In both cases the centre, the control authority, is transferring power downwards in order to increase the effectiveness of policy and to reduce the risk of failure. The essential thinking is the same in both cases, namely that if people are committed to the objectives of the policy then they are less likely to undermine it since they damage their own interests by so doing. The movement away from a centralized, coercive approach to policy is probably the major change that is brought about by a commitment to sustainable development.

Summary

- Biologists recognize three levels or forms of biodiversity: ecosystem, species and genetic diversity. Different definitions can lead to different strategies for conservation and different problems of instrument choice. This chapter is concerned with species diversity.
- The threats to species and their habitats can be very diverse and biodiversity policy can thus involve a wide range of instruments and much of what has been said in earlier chapters is thus applicable to biodiversity policy.
- The new factor that arises is from the problem of irreversibility of species loss. Because of this governments committed to sustainability place a premium on the risk of policy failure.
- One way of reducing the risk of failure is to incorporate a safety margin in the setting of instruments. A control authority incorporates a safety margin by setting lower permitted discharges for pollution, lower permitted catches for fish stocks, etc. than it would do were it not concerned with conserving biodiversity.

- Safety margins have resource costs for two reasons: the authority has to monitor to tighter standards and the firms have to invest more, or forgo more output, in order to meet those standards.
- Some authors have suggested that the resource costs may be saved by setting higher penalties for breaching standards instead of tighter standards. I have termed these financial safety margins rather than physical safety margins.
- But the scope for financial safety margins is limited since, if penalties are seen as out of scale to the gravity of the offence, juries will not convict and judges will not impose the penalties set. Legal sanctions must be seen as reasonable if they are to be enforced.
- Instruments are set according to the economic conditions facing those who are controlled. When those conditions change the instrument settings must be changed also.
- If the agent discovers that he can profit by breaching the standard and accepting the penalty before the control authority, or before the control authority can alter the settings, this is a first mover problem. To protect biodiversity the control authority must follow a strategy that affords protection against the first mover problem.
- It might do seek to do this by investing in detection of first mover opportunities. However the agent is in the best position to perceive profitable opportunities to himself.
- The alternative is for the authority to choose instruments that afford protection against first mover advantage.
- Negative instruments, such as command and control, pollution taxes and tradable permits, do not offer protection against first mover advantage. Thus if the polluter discovers it is to his advantage to breach his consent conditions and pay the consequent fine, he does not ask the control authority to increase the penalty.
- Negative instruments are predicated by the polluter-pays principle. Hence the polluter-pays principle does not protect biodiversity against first mover advantage.
- Protection against first mover advantage is afforded where the agent has some property rights in biodiversity. Positive instruments, where the agent is paid for his contribution to conserve biodiversity, offer some prospect of protection against the first mover problem. If the agent is able to profit from wildlife without putting its survival at risk, e.g. by charging for viewing it, then risk of first mover exploitation is again reduced.
- Moral hazard arises when the control authority is unable to distinguish damage to biodiversity from natural events outside the agent's control from damage that results from the failure of the agent to carry out his responsibilities. In these circumstances the agent has an incentive to neglect his responsibilities and blame the resulting damage on natural events. Biodiversity needs protection against moral hazard as well as first mover problems.
- As with first mover problems, protection is achieved by giving the agent some property rights in biodiversity. Negative instruments again offer no protection.
- The point noted in Chapter 15 that to achieve sustainable development people need to be bonded to the objectives is thus reiterated.

Chapter 17

Global pollution policy

The last chapter discussed how the sustainability objective of conserving biodiversity places a premium on avoiding failures of environmental policy. This gave a new perspective to the choice of instruments and exposed the limitations of the coercive approach to environmental policy. A more cooperative approach to policy, with property rights shared between agents and the control authority, appears to be the direction of progress. The second sustainability objective, protecting the integrity of the global atmosphere, also encourages cooperation but in this case the cooperation is required between the governments of nation states.

The atmosphere is a global public good; it is necessary for the existence of life and its life-giving services are both non-rival and non-excludable. Damage to its integrity from the emissions of greenhouse gases and ozone-destroying gases is equally non-excludable and safeguarding its integrity therefore requires cooperation between countries. Biodiversity, of course, is also a global public good and safeguarding it also requires international cooperation. This aspect of the problem was not discussed in the last chapter but is encompassed by what is said in this chapter. In general a commitment to sustainable development causes us to look outward and view environmental problems from a global perspective. This raises a set of issues not so far considered, namely environmental policy in situations where the environmental problems cross national boundaries. These issues are the subject of this chapter and they are considered in the context of global atmospheric pollution.

Trans-boundary pollution can be classified broadly into three categories:

1. unidirectional trans-boundary pollution;
2. regional reciprocal pollution;
3. global pollution.

I consider these in sequence.

Unidirectional trans-boundary pollution

Sometimes referred to as the 'upstream–downstream' problem, this is the simplest case of trans-boundary pollution and exists where economic activity in one set of nations conveys a negative externality (pollutes the environment of another). European examples are on the rivers Rhine and Danube, where activities by upstream countries both pollute the river and reduce the volume of flow and thereby cause problems for countries located downstream.

The simplest case of unidirectional trans-boundary pollution involves just two countries, one that is the polluter and one the sufferer. I consider this case first before considering what additional complications arise from there being more than one polluter and sufferer.

The new consideration that arises in this case is that, with separate sovereign countries, there is no control authority with powers to enforce a solution by the use of command and control or economic instruments. Settlement has to be by international agreement and with the consent of both parties. Agreement has to be reached through bargaining between the parties and the literature on trans-boundary pollution is largely concerned with analysis of bargaining. Bargaining problems are normally analysed through games theory. This can involve some complex mathematics and here I do no more than introduce some very basic concepts to illustrate the problems.

Trans-boundary pollution is assumed to be an externality in the sense of the definition used in Chapter 4, i.e. that there is a social surplus to be realized from reduction in it. This means that were pollution reduced, the gainers could compensate the losers and still be better off. In the context of trans-boundary pollution this means that there exists some reduction of pollution that the sufferer would value at a greater sum than it would cost the polluter to achieve it. About this proposition note two things:

1. There could be cases of trans-boundary pollution where this is not the case; where the economic costs to the polluter of reducing the pollution exceed the economic benefits to the sufferer from the reduction of the pollution. In the case of pollution occurring within national boundaries this would be where the existing degree of pollution was optimal.
2. The economic benefits (and indeed the costs) are those defined by the country concerned. If the pollution damages wildlife in the downstream country but that country is indifferent to wildlife then this damage does not feature in its calculations of the benefits of pollution reduction. On the other hand, if the country does value wildlife then it features in the issue. The critique of the Pigovian theory of pollution in Chapter 5 in a sense does not apply to trans-boundary pollution, since the sovereign state is the sole arbiter of its costs and benefits. But of course one major purpose of sustainable development and of the Earth Summit is to raise the profile of environmental issues, and to broaden the perspective of them, in countries that have other priorities.

If we assume that the treatment of trans-boundary pollution can generate a social surplus then bargaining about it is said to be a *positive sum game*. For the simple case of two-country unidirectional pollution the 'game' is for the polluter, A, to reduce pollution in return for compensatory payment from B. There is a solution to this game that will make both players better off than they presently are. The situation is portrayed in Fig. 17.1. P^* is the optimum

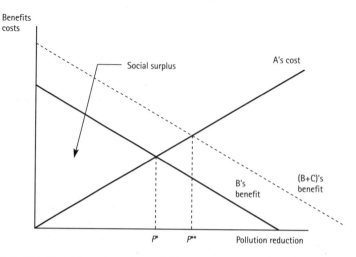

Fig. 17.1 Unidirectional trans-boundary pollution.

amount of pollution reduction and the triangle of the social surplus is shown. Payment of any part of this surplus from B to A in return for A reducing pollution to P^* will leave both parties better off.

But to reach this solution the parties have to cooperate. This requires mutual trust. A has to believe that, once it has reduced pollution, B will continue with the payments; B has to believe that if it makes the payments A will not subsequently threaten to increase the pollution again unless it receives greater payments. If this degree of trust does not exist (and note that there is no authority to enforce the agreement except with mutual consent) then a situation arises known as the *prisoner's dilemma* (Box 17.1), where both parties suffer from the lack of mutual trust. The prisoner's dilemma characterizes many potential international conflicts.

Bargaining between the parties will also be subject to other problems as well as mutual trust. Each has an incentive to exaggerate its position in order to strengthen its bargaining position. Thus A has an incentive to exaggerate the costs of pollution control and B to exaggerate the damage it sustains as a result of that pollution. While economists have devised mechanisms under which it will benefit countries to tell the truth in international relations, there is no authority to enforce any set of rules that ensure truth telling. Countries in any event do not have to deliberately lie. The issue will probably have reached the point of negotiation as a result of lobbying by interested parties in the countries concerned. They will also have an incentive to exaggerate their concerns: on the costs of pollution control or its benefits. All the country has to do is to faithfully reflect those views or to emphasize the views that enhance its bargaining position and play down those which weaken it. Non-governmental agencies concerned with environmental protection have learnt these rules of international bargaining and have sought to present their views directly in international fora. As a consequence of the tactical presentation of the costs

Box 17.1 The prisoner's dilemma

Two prisoners are arrested for a crime which they have committed. There is no evidence against them and the police have to rely on confessions. The prisoners are held in separate rooms and each is presented with the following information. If he confesses he will receive a sentence of 5 years. If he does not confess but his fellow does then he will receive a harsher sentence of 10 years. The options each faces are as follows:

Self\other	Confess	Not confess
Confess	5	5
Not confess	10	0

His safe strategy is to confess thereby giving himself a sentence of 5 years. If he does not confess and his fellow prisoner also does not confess then he will get off. But he cannot guarantee that his fellow prisoner will not confess and if he does then the sentence is doubled to 10 years. Thus both prisoners confess and are sentenced. Were they able to rely on each other they could have got off. The inability of each to guarantee the behaviour of the other has made them both worse off.

and benefits of trans-boundary pollution control, even if cooperation is achieved the outcome may well differ from the optimum P^*.

Where there are more than two countries involved in the problem because there are several upstream polluters and/or downstream sufferers, a further constraint on solutions arises because of incentives to free ride. Thus if one downstream country negotiates an agreement, all downstream countries benefit and those which are not parties to the agreement avoid payment of compensation. The result of free riding could be a suboptimal agreement or none at all. A suboptimal result is illustrated in Fig. 17.1. The dotted line is the benefits of pollution control when a third country, C, benefits as well as B. In this case more pollution reduction is warranted, an amount P^{**}. But if C free rides then the most that can be achieved is P^*. Fig. 17.2 shows a case where free riding leads to no agreement at all. B and C both have the same level of benefit but the benefit of each country alone lies below A's cost of pollution control so that no agreement on pollution control is possible. However the joint benefit of B + C will allow some pollution reduction with the optimum at P^*. Free riding by either B or C means that no agreement can be reached.

One final obstacle to agreement in the case of unidirectional agreement is a reluctance on the part of countries to agree to financial transfers of the form envisaged here. Such transfers are in conflict with the principles for the discussion of international negotiations on trans-boundary pollution agreed by the signatory countries of the Organization for Economic Co-operation and Development (OECD). These uphold the polluter-pays principle (PPP) as the basis for agreement. This is interpreted to mean that the polluter pays the costs

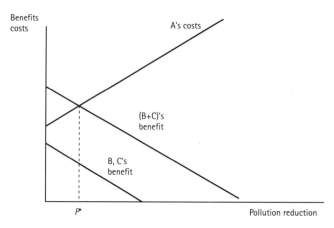

Fig. 17.2 Free riding and unidirectional pollution.

of pollution control and the sufferer bears the cost of any damage from the pollution that remains after an agreement has been reached. Our example in fact involves the opposite, known as the victim-pays principle (VPP). The reason for the VPP in this instance is that without it there is no basis for agreement since the polluter is made worse off by any pollution reduction and has no incentive to negotiate at all. PPP can thus serve to reduce the chances of agreement in the case of unidirectional trans-boundary pollution but it has a place in negotiations on other types of trans-boundary pollution, which now follow.

Regional reciprocal externalities

These exist when there is a set of countries that are simultaneously both the producers and the victims of trans-boundary pollution. The classic case is that of acid rain (precipitation of acids, principally sulphur dioxide, nitrogen oxides and hydrogen chloride) resulting from the combustion of fossil fuels, which both pollute the environment of the producing country but are also exported downwind to other countries. There is unlikely to be exact balance in regional reciprocal externalities; some countries will export more than they import and others will be in the reverse position. With acid rain, the UK exports far more than it imports; Scandinavian countries import far more than they export.

Since each country is to a degree polluting itself, some control will be exercised in the absence of agreement. This non-cooperative outcome is known as a Nash equilibrium and is shown in Fig. 17.3. The diagram is for one of the countries, B. Diagrams identical in form, but differing in the position and slopes of the private cost and benefit curves, will exist also for the others. The marginal benefit curve for pollution reduction by country B is designated as B_B and its cost curve is simply marked 'Costs'. The non-cooperative or Nash equilibrium, N, is where B_B is equal to the cost curve since this is the amount of pollution that B will wish to abate in its own interest. Any further reduction in pollution beyond N will cost country B more to achieve than it receives in benefit.

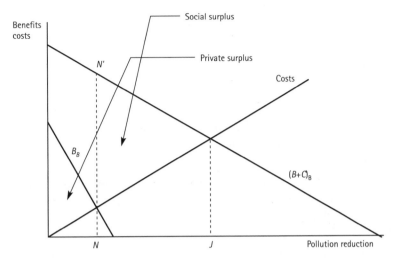

Fig. 17.3 Reciprocal pollution: cooperative and non-cooperative equilibria.

If country B adopts the Nash solution, its action will convey some benefit on other countries since B will produce and export less pollution. The joint benefit of all countries from pollution reduction is labelled $(B + C)_B$ and, as a result of B adopting the Nash solution, the international community will be at N' where the vertical dotted line intersects $(B + C)_B$. This is obviously not the optimum position. The optimum is the cooperative solution J where country B reduces its pollution to the point where its cost of control is equal to $(B + C)_B$. At J the social surplus, the net benefit to all countries from pollution reduction by country B, is at a maximum. If each country controls pollution to the point where its costs of control equal $(B + C)_B$ then all gain.

With regional reciprocal externalities all parties can hope to gain something from an agreement and a cooperative solution may be achievable without the need for payments between nations. The prospects for agreement are enhanced when there are no great disparities in pollution flows between the countries (i.e. exports and imports of pollution are similar in magnitude), since in this case no country has to incur great costs for little benefit nor can it receive great benefit for little cost. That is not the situation with the problem of acid rain, where the UK is in the former camp and Scandinavian countries in the latter. In consequence the UK initially declined to enter into the First Acid Rain Convention. The situation of the country illustrated in Fig. 17.3 is akin to that of the UK with acid rain. Its reductions in pollution convey great benefits on other countries as shown by the fact that $(B + C)_B$ is well to the right (above) B_B. It also receives little benefit from the reduction of others as shown by the fact that the Nash solution almost exhausts the benefit it derives from pollution reduction (only the small triangle below B_B to the left of the dotted line remains as the benefit from cooperation). Thus it has little incentive to cooperate whilst others have a strong incentive to persuade it to cooperate. There is a strong case therefore for the others to pay country B to cooperate.

They can do this either by money transfers of some of their gains to offset some of B's costs, or by benefit in kind, such as cooperation on some other international issue where B stands to gain more.

Global externalities

Global externalities can be viewed as a special case of regional reciprocal externalities where the region is the entire world and all countries stand to gain from cooperation. Global externalities are at the heart of sustainable development. The main ones are:

1. damage to the stratospheric ozone layer from the emission of chlorofluoro-carbons (CFCs) and halogens;
2. global warming; and
3. loss of biodiversity.

International action on each of these issues is the subject of international treaties. The Montreal Protocol commits signatories to phasing out the use of ozone-depleting chemicals. The Earth Summit resulted in the Framework Convention on Global Climate Change and the Convention on Biodiversity.

But even though all countries gain from controlling global pollution the gains are not necessarily evenly spread. This can be illustrated with global climate change from greenhouse gas emissions.

Global warming is predicted to lead to rises in sea levels and to shifts in the distribution of climatic zones. Its impact, and the costs resulting from it, will be very unevenly distributed across the world. Maritime states are the ones affected directly by sea-level rises and those with large areas of low-lying land, such as Bangladesh and The Netherlands, stand to lose most. Some low-lying islands of the Indian Ocean are predicted to disappear. Shifts in climatic zones seem likely to benefit sub-Arctic territories where cropping will become possible, but the arid zone could also spread northwards and affect the Mediterranean basin and the southern USA.

Equally the costs of curbing global pollution may not be evenly distributed either. The costs of any strategy to counteract global warming fall heavily in the short run on Western developed countries, particularly the USA, which are the major consumers of fossil fuels. In the longer run, however, control impacts on the Third World countries, which are denied the energy-intensive path to economic development followed in the West.

Since the benefits from global warming control are only experienced in the longer term and the costs are substantial, serious cooperative action on global climate change is problematic.

With biodiversity, while the benefits yielded by the exploitation of genetic capital in the form of new drugs, strains of food plants, etc. are ultimately widely distributed, the products are developed by companies, usually multinationals based in Western countries, who receive most of the profits. The countries from which the genetic material derives, where the plants and animals that are the sources of the discoveries are to be found, have in the past rarely received an adequate share of the profits. Species diversity reaches its maximum in tropical wet forests and, without the cooperation of the countries in which these critical reservoirs of genetic capital are found, a biodiversity convention would be ineffective. But these are developing countries and the tropical forests represent their resource base for development.

With both global warming and biodiversity the prospects of effective cooperation are limited by concerns over the distribution of costs and benefits between participants and in both cases also resource transfers from Western countries to the Third World are needed to obtain the latter's cooperation.

It might be thought that the ozone layer is different. If it is destroyed, life on earth would be impossible; this clearly affects all countries equally. But if the benefits of protecting the ozone layer are evenly spread, the costs of action are not, and parties have an incentive to free ride to in order to redistribute the burden of costs. Protecting the ozone layer entails replacing CFCs in refrigerators and air-conditioning systems with more expensive alternatives. This will place heavy costs on Third World countries who are in the process of equipping their populations with these goods. Western countries can phase in the new equipment as old, ozone-unfriendly equipment wears out. But it is also a small number of Western-based companies who are developing the alternatives to CFCs. Thus the profits from phasing out CFCs will offset the costs. The Third World has therefore sought financial transfers from the West to offset this expense.

Summary

- The global atmosphere and biodiversity are both global public goods and the sustainability objective of safeguarding the integrity of the atmosphere and conserving biodiversity require action on a global scale.
- The economics relevant to these problems is that concerning trans-boundary pollution. This is concerned with the control of pollution and other environmental problems that cross international boundaries.
- With trans-boundary pollution there is no control authority to operate command and control or economic instruments and action requires cooperative agreement.
- Trans-boundary pollution is classified into three types: unidirectional trans-boundary pollution; regional reciprocal trans-boundary pollution; and global pollution. The last is a special case of the second.
- Trans-boundary pollution exists if the suffering countries assess their benefits from reducing the pollution at more than the polluting countries assess their costs from controlling that pollution. In this case there is a potential Pareto improvement from reducing the pollution. The gainers could compensate the losers and still be better off.
- The countries concerned are the sole arbiters of the respective costs and benefits and it is their valuations that count. If they do not value some aspects of the environment such as wildlife then benefits to wildlife will not feature in the design of solutions. One objective of the Earth Summit is to raise the profile of environmental issues in participating countries. In so far as it succeeds, it increases recognized problems of trans-boundary pollution.
- Even with the simplest case of unidirectional trans-boundary pollution there are obstacles to realizing the social surplus, namely that parties have an incentive to exaggerate the costs and benefits of control, cooperation requires mutual trust that agreements will be kept to and where there is more than one polluter or more than one sufferer there are incentives to free ride.

- The lack of trust constitutes the prisoner's dilemma. It may serve to prevent a solution from which all will benefit. Free riding may mean that any agreement reached is suboptimal.
- Realizing the social surplus requires the victim-pays principle that the victim who gains from the reduction of pollution pays the polluter, who incurs the costs of control. By agreement, OECD countries are committed to the polluter-pays principle and adherence to this may pose an obstacle to agreement.
- But the payments by the victim need not take the form of money transfers. They may be in kind such as cooperation on some other issue where the status of victim and polluter are reversed.
- With regional reciprocal externalities all countries are both victims and polluters. Hence some abatement will occur in the absence of agreement. The non-cooperative solution is known as the Nash equilibrium. The Nash solution leads to less than the optimum degree of pollution abatement. There remain additional benefits from cooperation.
- A cooperative solution to a regional reciprocal externality is most likely if all countries stand to gain considerably from cooperating. This will be the case where exports and imports of pollution are more or less in balance for all parties. This is not the situation with acid rain.
- It is not the case also with global problems. With global climate change the costs and benefits are very unevenly distributed. With destruction of the ozone layer the benefits of a solution are evenly spread in the long term since the ozone layer is necessary to sustain life but the costs of control are unevenly distributed between the Third World and the West. With conservation of biodiversity also the costs of control fall largely on the Third World.
- In all of these cases, effective agreement requires money transfers between the West and the Third World and this factor serves to inhibit cooperation.

Chapter 18

Retrospect: sustainability, economics and the environment

Sustainable development is, by intention, a process of reorientation of individual economic and social behaviour on a global scale in order to safeguard the interests of future generations. As such it transcends the scope of economics although, along with the physical and biological sciences and other social science disciplines, economics has a role to play in it.

In this book I have tried to explain the contribution of economics. It is an important one. Economic analysis is required to reveal the resource implications and the implications for social welfare, of alternative courses of action. It is necessary also for designing and evaluating instruments to achieve specific purposes. Economics cannot determine those choices: it can only provide information on the implications of alternatives. Specifically it can say something about their opportunity costs. But it cannot express all environmental effects and consequences as meaningful monetary values, nor does it possess a simple formula for resolving the fundamental social conflicts that sustainability gives rise to. Cost–benefit analysis is a useful tool of analysis; it is not the philosopher's stone.

I finish the book by commenting on two issues of sustainability that have been mentioned in passing but on which more needs to be said, since they are at the centre of the sustainability debate. These are the issues of equity and population growth.

I have argued in earlier chapters that the burden of the costs of sustainability needs to be fairly distributed if the commitment to sustainable development is to be accepted; if there is not widespread acceptance then sustainable development is unattainable, since neither on a national scale, nor still less internationally, can it be imposed on unwilling citizens from the centre.

On a national scale, policy instruments to achieve sustainability face two potential problems:

1. a build-up of consumer resistance to environmental improvements, which are perceived as both excessively expensive and as being unfairly imposed;
2. the 'green consumer syndrome'; a belief that one has only to buy green products and thus to pay the appropriate green taxes and the environmental crisis will be solved.

While these two problems seem in conflict, they can coexist. As an example, consider a green tax on petrol to reduce CO_2 emissions. The richer sections of the community, who by and large do not pay for their motoring anyway, will absorb the tax and those on the margins of affording motoring will bear the brunt of the reduction of emissions. The rich may manifest the green consumer syndrome, not modifying their behaviour because they are being green in paying the tax. Those who pay the tax will be poorer and manifest resentment that motoring is being put out of their reach by taxation which has no effect on the rich.

Economists are practised at calculating the distributional consequences of different types of taxes and subsidies, and economics has long recognized that efficiency and equity can conflict in social policy. A tax on a household's use of energy may give the right price signals and encourage people to save energy, but it might place an excessive burden on the poor. A subsidy for insulation to reduce energy use may be inefficient but will benefit the poor who otherwise cannot afford to insulate their houses and reduce their fuel bills.

The international aspect of equity relates to the differences between the developed countries of the West and the Third World. In consequence it is compounded with the problem of population growth, since a short-hand description of the environmental crisis might be that it is the result of over-consumption in developed countries and excessive population growth in the Third World. This dilemma can be demonstrated by looking at the critical issue of the consumption of energy. Table 18.1 provides the basic information.

A 30% reduction of per capita energy consumption in the developed world, the sort of reduction that some commentators have argued is necessary as part of a strategy to control global warming, would allow Third World consumption at current population levels to increase by 66%. This would still only be 30% of (reduced) developed world levels. Alternatively this reduction would service 5.9 years of Third World population growth at current levels of per capita consumption.

Table 18.1 Energy saving, population and population growth (1986 data)

	Developed world*	Third World†
Population (millions)	1065	3852
Population growth rate (% per annum)	0.67	1.94
Energy consumption per capita (kg coal equivalent)	5621	710
Total energy consumption (percentage of world total)	73	27

* North America, Europe, Oceania, (ex) USSR.
† Latin America, Africa, Asia.
Source: UN Year Books.

To equalize per capita consumption at 1986 population levels, developed world consumption levels would have to fall by 81%. This would furnish 14.7 years of Third World population growth at existing consumption levels. Thus high population growth compounds the problems of international equity in moving to sustainable development.

Whether in the face of this arithmetic the mitigation of global warming is politically feasible is debatable. Certainly the requirements for resource-saving technology are severe. The message that the developed world is sending out is that developing countries may not exploit their natural resources in the way that the West has exploited its resources, since to do so will bring global disaster. The Third World is rightly demanding that Western countries, which consume the bulk of resources and produce the bulk of the pollution, should bear the main burden of adjustment.

Within the West it is the wealthier portion of the population that consumes the major part of the natural resources. If the adjustments to overcome the environmental crisis are to be made, it is important that the distribution of the sacrifices is accepted as equitable. This means that the incidence must fall more heavily on the rich than the poor. Market measures will not ensure this; rather they tend to operate within the existing structure of market power.

Further reading: a brief guide

General

The neoclassical approach to sustainability and environmental economics is captured in the set of Blueprints produced by Earthscan: David Pearce, Anil Markandya and Edward B. Barbier (1989) *Blueprint for a Green Economy*; David Pearce (ed.) (1991) *Blueprint 2. Greening the World Economy*; David Pearce (ed.) (1993) *Blueprint 3. Measuring Sustainable Development*. Textbooks in the neoclassical mould are David W. Pearce and R. Kerry Turner (1990) *Economics of Natural Resources and the Environment*, Harvester Wheatsheaf; and Tom Tietenberg (1994) *Environmental Economics and Policy*, Harper Collins. Robert Dorfman and Nancy S. Dorfman (eds.) (1993) *Economics of the Environment. Selected Readings*, 3rd edn. W.W. Norton & Co. contains some of the classic material.

Economic preliminaries

There are a large number of microeconomics textbooks that cover the material of this section. They differ in the degree of difficulty and in the extent to which they cover material not discussed in this book.

The choice of instruments for environmental policy

The classic text is William J. Baumol and Wallace E. Oates (1988) *The Theory of Environmental Policy*, 2nd edn, Cambridge University Press. This however is hard going and involves a moderate degree of mathematics. Readings in Dorfman and Dorfman op. cit. are more accessible. John Pezzey (1988) Market mechanisms of pollution control: 'polluter pays', economic and practical aspects', in R. Kerry Turner (ed.) *Sustainable Environmental Management. Principles and Practice*, Belhaven Press, raises some interesting aspects of the practical problems of instrument choice for pollution control. An antidote to the neoclassical views on pollution control is Mikael Skou Anderson (1994) *Governance by Green Taxes. Making Pollution Prevention Pay*, Manchester University Press. Nitrate pollution in the UK is covered by Nick Hanley (1991) The economics of nitrate pollution control in the UK, in N. Hanley (ed.) *Farming and the Countryside: an Economic Analysis of External Costs and Benefits*, CAB International. The references given in the chapters are recommended for household waste and air pollution by road traffic. On the latter, the Royal Commission Report is authoritative.

Cost–benefit analysis

There are many texts on cost–benefit analysis of varying technical difficulty. Despite its age the best introductory text remains E.J. Mishan (1971) *Cost Benefit Analysis: an Informal Introduction*, George Allen & Unwin. Most of the cost–benefit analysis texts contain sections on valuing the environment or intangibles in general. In addition there are numerous works on the specific issue of environmental valuation from the viewpoint of environmental economics. Since it is a central issue to the neoclassical approach most of the textbooks devote considerable space to it. David Pearce (1993) *Economic Values and the Natural World*, Earthscan is an accessible exposition of the approach that is criticized in this book.

Sustainable development

The literature on sustainable development is growing almost too rapidly to cope with. The Blueprint series covers the views criticized here. Chapters 1–3 of David Pearce, Edward Barbier and Anil Markandya (1990) *Sustainable Development: Economics and Environment in the Third World*, Earthscan cover the idea of aggregate natural capital. In addition three disparate works are recommended: Michael Redclift (1987) *Sustainable Development. Exploring the Contradictions*, Routledge; Michael Common (1995) *Sustainability and Policy. Limits to Economics*, Cambridge University Press; and Michael Jacobs (1991) *The Green Economy. Environment, Sustainable Development and the Politics of the Future*, Pluto Press. Exhaustible and renewable resources are covered in most of the texts. On the issue of international agreements, the article by Scott Barrett (1990) The problem of global environmental protection. *Oxford Review of Economic Policy* **6**(1), and reprinted in several places including (under a slightly different name) in Dorfman and Dorfman op. cit., covers the general problems of cooperation. Karl Göran-Mäler's article, 'International environmental problems' which again is available in a number of texts including Dieter Helm (ed.) (1991) *Economic Policy Towards the Environment*, Blackwell, is mainly concerned with acid rain. Alice Enders and Amelia Porges (1992) Successful conventions and conventional success: saving the ozone layer, in K. Anderson and R. Blackhurst (eds.) *The Greening of World Trade Issues*, Harvester Wheatsheaf, is as its name suggests. There are many texts on the issue of global climate change and the scene is changing rapidly. William R. Cline (1992) *The Economics of Global Warming*, Institute of International Economics, Washington DC, is a good survey of the issues. On the issue of biodiversity, two publications of international organizations, OECD Environment Directorate (1993) *Economic Incentives for the Conservation of Biodiversity: Conceptual Framework and Guidelines for Case Studies* and Theodore Panayotou (1994) *Economic Instruments for Environmental Management and Sustainable Development*, UN Environmental Programme, Expert Group on the Use and Application of Economic Policy Instruments for Environmental Management and Sustainable Development, survey instruments for biodiversity, although it is not clear from either what distinguishes biodiversity from other environmental problems. Academic economists are largely concerned with other problems. The issue of the measurement of biodiversity and the significance of it for human society are unresolved.

References

Arrow, K. J. and Fisher, A.C. (1974) Environmental preservation, uncertainty and irreversibility. *Quarterly Journal of Economics* **88**, 312–319.

Ball, S. and Bell, S. (1991) *Environmental Law*. Blackstone Press.

Barnett, H. J. (1979) Scarcity and growth revisited, in V. Kerry Smith (ed.) *Scarcity and Growth Revisited*. Baltimore: Resources for the Future, pp. 163–217.

Barnett, H. J. and Morse, C. (1963) *Scarcity and Growth*. Baltimore: Resources for the Future.

Bowers, J. (1979) Do we need more forests? School of Economic Studies Discussion Paper No. 103, University of Leeds.

Bowers, J. (1988) Cost–benefit analysis in theory and practice: agricultural land drainage projects, in R. Kerry Turner (ed.) *Sustainable Environmental Management. Principles and Practice*. London: Belhaven Press.

Bowers, J. (1991) The economics of peat extraction, in Commission of Inquiry into Peat and Peatlands. *Public Hearing No. 3. Plantlife*. London: Natural History Museum.

Bowers, J. (1993) Pricing the environment: a critique. *International Review of Applied Economics* **7**(1), 91–107.

Bowers, J. and Hopkinson, P. (1994) The treatment of landscape in project appraisal: consumption and sustainability approaches. *Project Appraisal* **9**(2), 110–118.

Clawson, M. (1959) *Methods of Measuring the Demand for and Value of Outdoor Recreation*. Washington DC: Resources for the Future.

Commission of Inquiry into Peat and Peatlands (1991). *Public Hearing No. 3. Plantlife*. London: Natural History Museum.

Commission on the Third London Airport (1971) *Report*. London: HMSO.

Dalvi, M.Q. and Nash C.A. (1977) The distributive impact of road investment, in P. Bonsall, P.J. Hills and M.Q. Dalvi (eds) *Urban Transportation Planning*. London: Abacus Press.

Department of the Environment (1990) *This Common Inheritance*. London: HMSO.

Department of the Environment (1992) *The UK Environment*. London: HMSO.

Department of the Environment and Welsh Office (1995). *A Waste Strategy for England and Wales, Consultation Draft*. London: HMSO.

Department of Transport (1989) *Roads to Prosperity*. London: HMSO.

Department of Transport (1994) Standing Advisory Committee on Trunk Roads Appraisal (SACTRA). *Trunk Roads and the Generation of Traffic*. London: HMSO.

Ecos (1996) vol. 17, no. 1. Symposium on Local Agenda 21.

Environmental Resources Ltd (1992) *Economic Instruments and Recovery of Resources from Waste*. Department of Trade and Industry and Department of the Environment London: HMSO.

Fisher, A.C. (1981) *Resource and Environmental Economics*. Cambridge: Cambridge University Press.

Goodwin, P. (1992) A review of new demand elasticities with special reference to the short and long run effects of price changes. *Journal of Transport Economics and Policy*, **26**, 155–169.

Hawksworth, D.L. (ed.) (1994) *Biodiversity. Measurement and Estimation*. London: Chapman & Hall.

Helm, D. and Pearce, D. (1990) Assessment: economic policy towards the environment. *Oxford Review of Economic Policy* **6**(1).

Henry, C. (1974) Investment decisions under uncertainty: the irreversibility effect. *American Economic Review* **64**, 1006–1012.

Johansson, P.-O. (1990) Valuing environmental damage, *Oxford Review of Economic Policy* **6**(1).

Mandeville, B. de (1924) *The Fable of the Bees: or Private Vices, Publick Benefits*. Oxford: Oxford University Press.

Meadows, D.H., Meadows, D.L., Randers, J. and Behrens, W.W. III (1972) *The Limits to Growth. A Report of the Club of Rome's Project on the Predicament of Mankind*. Washington DC: Potomac Associates Inc.

Pearce, D.W. and Nash C.A. (1981) *The Social Appraisal of Projects*. London: Macmillan.

Pearce, D., Markandya, A. and Barbier, E.B. (1989) *Blueprint for a Green Economy*. London: Earthscan.

Pearce, D., Barbier, E.B. and Markandya, A. (1990) *Sustainable Development. Economics and Environment in the Third World*. London: Earthscan.

Pigou, A.C. (1920) *The Economics of Welfare*. London: Macmillan and Co.

Repetto, R. (1985) *Paying the Price – Pesticide Subsidies in Developing Countries*. Research Report 2. Washington: World Resources Institute.

Repetto, R. (1986) *Skimming the Water – Rent-seeking and the Performance of Public Irrigation Systems*. Research Report 4. Washington: World Resources Institute.

Repetto, R. (1988) *The Wood for the Trees – Government Policies and the Misuse of Forest Resources*. Washington: World Resources Institute.

Rico, R. (1995) The US allowance trading system for sulfur dioxide: an update on market experience. *Environmental and Resource Economics* **5**(2), 115–129.

Royal Commission on Environmental Pollution (1994). Eighteenth Report. *Transport and the Environment*. London: HMSO.

Smith, V. Kerry, (ed.) (1979) *Scarcity and Growth Revisited*. Baltimore: Resources for the Future.

Sweet, J. (1994) Critical loads, the new sulphur protocol and European energy policy. *European Environment* **4**(1), 2–4.

Turner, R. Kerry (ed.) (1988) *Sustainable Environmental Management. Principles and Practice*. London: Belhaven Press.

Ulph, A. and Ulph, D. (1994) Global warming: why irreversibility may not require lower current emissions of greenhouse gases. Discussion Papers in Economics and Econometrics No. 9402, University of Southampton, Department of Economics.

World Commission on Economic Development (1987) *Our Common Future*. London: Oxford University Press.

Young, M.D. (1992) *Sustainable Investment and Resource Use: Equity, Environmental Integrity and Economic Efficiency*. UNESCO.

Index

Page numbers in **bold** refer to definitions